CLOUDS, RAIN AND RAINMAKING

CLOUDS, RAIN AND RAINMAKING

B. J. MASON

Director-General, Meteorological Office, Bracknell, England

SECOND EDITION

CAMBRIDGE UNIVERSITY PRESS

CAMBRIDGE

LONDON · NEW YORK · MELBOURNE

CAMBRIDGE UNIVERSITY PRESS
Cambridge, New York, Melbourne, Madrid, Cape Town, Singapore,
São Paulo, Delhi, Dubai, Tokyo, Mexico City

Cambridge University Press
The Edinburgh Building, Cambridge CB2 8RU, UK

Published in the United States of America by Cambridge University Press, New York

www.cambridge.org
Information on this title: www.cambridge.org/9780521157407

First published 1962
Second edition 1975
First paperback edition 2010

A catalogue record for this publication is available from the British Library

Library of Congress catalogue card number: 74-16991

ISBN 978-0-521-20650-1 Hardback
ISBN 978-0-521-15740-7 Paperback

CONTENTS

PREFACE TO THE SECOND EDITION

This book has the same structure and scope as the first edition, but the text has been revised to accommodate the principal advances that have taken place during the intervening decade. Again it is addressed to a wide range of readers who may wish to have a concise, readable and up-to-date account of cloud physics. More detailed treatments of all the topics and original references will be found in the new edition of the author's *The Physics of Clouds*.

I am greatly indebted to Miss Eileen Forde for her invaluable help in preparing the manuscript and index.

B. J. M.

METEOROLOGICAL OFFICE
BRACKNELL
March 1974

ABRIDGED PREFACE TO FIRST EDITION

In this small volume I have attempted to write a concise, up-to-date account of recent researches on the formation of clouds and the development inside them of rain, snow, hail and lightning. I hope that it will interest a wide range of readers who may wish to know rather more about cloud physics than may be gleaned from popular articles but do not require the detailed treatment given in the author's *The Physics of Clouds*.

Although much of the text requires little more than a good knowledge of elementary physics, enough mathematics is included to indicate the kind of calculations that can be made and the magnitudes of the quantities involved. The do-it-yourself experiments described at the ends of chapters are designed to encourage the reader to observe some of the phenomena at first hand.

I shall be pleased if this book does something to enhance the enjoyment and understanding of those who watch the sky and also reveals the pleasure to be gained from studying a wide range of physical processes both in the atmosphere and the laboratory.

For permission to reproduce many of the diagrams I am indebted to those learned societies, publishers, and authors acknowledged in the text.

I am particularly grateful to Dr G. Haigh for checking the proofs, to Miss S. H. Wood for typing the manuscript and to Miss Y. Banos for her help in preparing the Index.

1

CLOUD FORMS AND FEATURES

Introduction

Clouds are formed when air containing water vapour rises, expands under the lower pressures which exist at higher levels in the atmosphere, and thereby cools until some of the vapour condenses into a cloud composed of myriads of tiny water droplets.

At first glance a cloudy sky may appear chaotic, but the perceptive observer will discern some semblance of order, the existence of recognizable patterns and of distinctive cloud types, all of which, in their infinite variety of shape and form, are somehow expressions of the way in which the air has risen to fashion them.

The meteorologist classifies clouds mainly by their appearance according to an international system essentially similar to that originally proposed, 150 years ago, by Luke Howard, a London-born pharmacist. But because the dimensions, shape, structure and texture of clouds are largely influenced by the kind of air movements which result in their formation and growth, and by the behaviour of the cloud particles, much of what was originally a purely visual classification can now be justified on physical grounds. The main cloud types are described in table 1.

Sometimes the clouds form a vast sheet hundreds of miles across, indicating a steady ascent of air over large areas; at other times they are scattered over the sky in isolated puffs and heaps revealing irregular local upcurrents with clear spaces in between. These are respectively the *stratiform* or layer clouds, and the *cumuliform* or heap clouds. With either kind the prolonged ascent of air leads to successively deeper and denser clouds, and finally to rain: widespread continuous rain from the layer clouds and showers from the heap clouds.

Storm clouds

The depressions or cyclones of temperate latitudes usually bring both kinds of rain, an individual storm bringing several hours of continuous rain followed by showers and brighter intervals. It is heralded by the appearance, at heights of more than 20000 ft

TABLE 1. *Principal classes of clouds*

I. Stratiform or layer clouds

Cloud type	Abbreviation	Description	In temperate regions		Type of vertical air motion
			Height range	Temp. range	
High-level clouds					
Cirrus	Ci.	Detached clouds composed of delicate white filaments and appearing in either tufts, streaks, trails, feather plumes, or bands	Above 20000 ft	Below −25 °C	Widespread, prolonged and regular ascent with vertical velocities of typically 5–10 cm/s
Cirrocumulus	Cc.	A dappled layer or patch of cloud forming amongst cirrus. Composed of small white flakes or very small globules arranged more or less regularly in groups or lines, or more often as ripples resembling those of sand on the sea shore			
Cirrostratus	Cs.	A fused sheet of cirrus cloud which does not obscure the sun or moon, but gives rise to halos Sometimes it appears as a diffuse white veil across the sky			
Medium-level clouds					
Altostratus	As.	A grey, uniform, striated or fibrous sheet but without halo phenomena, and through which the sun is seen only as a diffuse bright patch or not at all	7000–20000 ft	0 to −25 °C	
Altocumulus	Ac.	A dappled layer or patch of cloud composed of flattened globules which may be arranged in groups, lines or waves collectively known as billows			
Low-level clouds					
Stratocumulus	Sc.	A layer of patches composed of laminae or globular masses arranged in groups, lines or waves and having a soft, grey appearance. Very often the rolls are so close together that their edges join and give the undersurface a wavy character. Stratocumulus (cumulogenesis) is formed by the spreading out of the tops of cumulus clouds, the latter having disappeared	Below 7000 ft	Usually warmer than −5 °C	Widespread irregular stirring with vertical velocities usually less than 10 cm/s

Stratus	St.	A uniform, featureless layer of cloud resembling fog but not resting on the ground. When this very low layer is broken up into irregular shreds it is designated fractostratus	Usually within 1000 or 2000 ft of the ground	—	Widespread irregular stirring and lifting of a shallow layer of cool, damp air formed near the ground
Nimbostratus	Ns.	An amorphous, dark grey, rainy cloud reaching almost to the gound	As for St.		Widespread regular ascent with vertical velocities of 5–20 cm/s

Clouds with marked vertical development

II. Cumuliform or heap clouds

Cumulus	Cu.	Detached, dense, clouds with marked vertical development; the upper surface is dome-shaped with sharp-edged rounded protuberances, while the base is nearly horizontal	Extend from 2000 to 20000 ft or more	—	Convective motion in which large bubbles of warm air rise with vertical speeds of 1–5 m/s
Cumulonimbus	Cb.	Heavy masses of dense cloud, with great vertical development, whose cumuliform summits rise in the forms of towers, the upper parts having a fibrous texture and often spreading out into the shape of an anvil. These clouds generally produce showers of rain and sometimes of snow, hail or soft hail, and often develop into thunderstorms	May extend up to 40000 ft	Summits may be as cold as – 50 °C	Strong convective motions with vertical upcurrents of 3 to more than 30 m/s

III. Special types of cloud

Fracto clouds: fractocumulus fractostratus fractonimbus		Fragments of low cloud which are associated with cumulus, stratus or nimbostratus as the case may be	—	—	Indeterminate
Castellanus	Ac-cas.	Miniature turreted heap clouds forming at medium levels usually in lines. In summer they are symptomatic of the approach of thundery weather	As for Ac.	—	Convective motions released at middle levels by the slow lifting of unstable air often ahead of cold fronts
Orographic clouds Lenticular and wave clouds		When air is forced to ascend a hill or mountain barrier, a smooth, lens-shaped cloud with well-defined edges may form over the summit. This is a lenticular cloud. If the air flow is set into oscillation by the hill, a succession of such clouds may form in the crests of the stationary waves produced in the lee of the mountain. These are designated wave clouds	—	—	The upcurrents in these clouds are usually quite strong—of order 1–10 m/s

(6 km) above the ground, of thin high clouds either in the form of trails or streaks composed of delicate white filaments, or as tenuous white patches and narrow bands. In recognition of their fibrous or hair-like texture, clearly seen in pl. I(a), they are called *cirrus** clouds. After approaching steadily for two or three hours, the leading cirrus arrive overhead and the barometer begins to fall. Gradually the clouds thicken, and the bands merge to form a great veil of silken cloud covering the whole sky. This sheet of cirrus cloud is called *cirrostratus*. At first it is translucent; the sun shines palely through it, and sometimes we may notice that centred on the sun but at an angular distance of 22° from it, there is a *halo*, a bright ring of light tinged with orange-red on the inside (see pl. I(b)). This and other beautiful optical phenomena are caused by the refraction of sunlight through the ice crystals formed at these levels, where the temperature is usually below −25 °C. This cirrostratus shield usually has an irregular texture so that only fragments of the halo are seen, but sometimes the cloud is very thin and diffuse, perhaps just a milky film over the blue sky, and then the halo may appear very bright perhaps with other rings and arcs around it, some with colours as bright as those of the rainbow.

Four or five hours after the appearance of the first cirrus, the clouds overhead have thickened and darkened so much that the halo is extinguished and the sun is seen only as a brighter patch with smudged edges. At this stage, depicted in pl. II(a), the cloud is termed *altostratus*; it continues to thicken, becomes lower, and patches of cloud form beneath it, making dark wavy patterns. With the closer approach of the storm centre, the wind freshens, the barometer falls rapidly, and within an hour or two the first raindrops fall. The rain soon becomes heavier and the cloud base lowers closer to the ground. It is difficult to discern any real structure in this grey *nimbostratus* or rain cloud, but beneath it tattered fragments of scud (*fractostratus*) are driven along in the wind. Exploration of this rain cloud with radar and aircraft reveals that it is usually composed of several thick layers. At a height of 2 or 3 km the rain gives way to snow, and higher still the snowflakes become smaller and are replaced by clouds of ice crystals which reach up to cirrus levels. In the regions of heavier snow and rain the gaps between the layers become filled in, so that the cloud extends upwards continuously from near the ground to a height of 8 or 10 km.

* Cirrus = lock of hair.

Plate I. (*a*) Cirrus cloud in rather dense, fibrous patches containing curved streaks (mare's tails) with hooked or tufted heads. These clouds are composed almost entirely of ice crystals. (Photograph from Clarke Collection; by permission of the Royal Meteorological Society.) (*b*) Cirrostratus. A thin, uniform white veil of ice-crystal cloud showing a halo around the sun. (Photograph from Clarke Collection; by permission of the Royal Meteorological Society.)

After some hours the steady rain ends and the barometer ceases to fall; the sky becomes lighter as much of the upper cloud becomes thin and broken or clears away, but the lower overcast of ragged *stratus* clouds persists and gives intermittent drizzle, whose drops are smaller but more numerous than raindrops.

Sometime later the wind freshens again, the sky darkens and there is heavy rain for a short while, quickly followed by a sudden improvement in the weather. The wind veers sharply, becomes gusty and cool, the barometer begins to rise, and the low clouds lift and break. Above them the dappled high clouds and cirrus recede to leave a clear sky.

The widely scattered low clouds are now *cumulus*, the heap clouds, which have fairly level bases and rounded tops like cauliflowers (see pl. II(*b*)). Some of these clouds grow larger and taller and, within about 15 min, tower up to 20000 ft (6 km) or more. Before long we see beneath the largest clouds descending trails of rain; they have become *cumulonimbus*, the shower clouds. Usually there is a striking change in the appearance of the cloud tops as the shower develops. The sharp, clear outlines of the cauliflower become smudged, ragged and soft; the bulges flatten, the cracks become filled in and the upper part of the cloud takes on the fibrous texture of cirrus. Often this is drawn sideways by the stronger winds aloft, projecting beyond the cloud base in the shape of an anvil as shown in pl. III(*a*). Eventually it may separate from the lower part of the cloud, which usually subsides and dissolves with the development of the shower, and drifts away as a slowly evaporating mass of 'anvil cirrus'.

An individual shower cloud may decay within half an hour of its inception, but as one tower releases its rain and evaporates, another springs up on its flanks, and the cloud mass may travel as a recognizable whole for several hours.

After some hours, convection declines; the larger clouds appear even more rarely, the showers cease and the wind moderates. The barometer continues to rise, but more slowly, and finally even the small cumulus disappear. The whole cloud system of the storm has passed away.

Some other forms of clouds

On the fringes of the large storm-cloud systems and in weaker storms the layer clouds are usually thinner and broken up into dapples or parallel rolls (billows). The top of a cloud layer tends

Plate II. (*a*) Altostratus. A thick, irregular layer of ice-crystal cloud with the sun shining palely through it. (Photograph from Clarke Collection; by permission of the Royal Meteorological Society.) (*b*) A line of cumulus clouds with fairly level bases and rounded, cauliflower-like tops. (Photograph from Clarke Collection; by permission of the Royal Meteorological Society.)

to cool by radiating heat into space, while the interception of the earth's radiation at the base of the layer tends to warm it. After some time slow convective motions are produced in the layer giving it a dappled structure. All shallow layer clouds soon assume this structure unless they are shielded by a higher cloud layer. The dapple clouds which form amongst cirrus are called *cirrocumulus*. These are composed of small white flakes which may be more or less regularly arranged in lines or in ripples resembling those of sand on the seashore. Uniformly distributed dapples occur when the wind is almost the same throughout the cloud layer, but when the wind changes in speed and/or in direction with height, the globules become arranged in lines or rolls. The dapple clouds are conventionally classified as *cirrocumulus*, *altocumulus* (pl. III(*b*)), or *stratocumulus* according to their height above ground—see table 1.

In stratocumulus layers, which may be quite thick, the rolls are often so close together that their edges join so that the undersurface has only an undulating appearance without any clear chinks. Such clouds sometimes cover the sky for days on end in spells of quiet winter weather.

All the dapple clouds, except perhaps some of the high cirrocumulus, are composed of liquid water droplets even though their temperature may be well below 0 °C, the ordinary freezing-point of water. (The occurrence of water drops in the *supercooled* state will be discussed in chapter 4.) In these clouds one commonly observes *coronae*, diffraction patterns of coloured rings around the sun, which are easily distinguished from halos because they are much closer to the sun, and usually have bright colours, with *red* on the outside. They are most distinct when the droplets are all of uniform size, otherwise the colours become mixed to produce a dull brown-red ring separated from the sun by a bluish white zone. Because of the glare, halos, coronae and other optical phenomena around the sun should be looked at through dark glasses, but they are usually much more splendid than the fainter ones more commonly noticed around the moon.

We have now described all the principal cloud forms, but the rather special varieties produced under the influence of orography deserve a separate word. When air ascends over only a small hill the vertical motions set up in the air may be appreciable even at cirrus levels. When they are sufficiently pronounced they produce clouds which

Plate III. (a) Cumulus and cumulonimbus. To the left and in the foreground are cumulus clouds with characteristic well defined edges. The larger cloud in the background has the ragged, fibrous edges which develop when the cloud is transformed into ice crystals. The anvil-shaped ice cloud is a marked feature of the cumulonimbus. (Photograph from Clarke Collection; by permission of the Royal Meteorological Society.) (b) Altocumulus. A medium-level cloud layer broken into a series of parallel rows or billows. (Photograph by R. S. Scorer.)

remain almost stationary relative to the hill although a strong wind may be blowing. The air stream is set into oscillation in being forced over the hill and the clouds form in the crests of the (almost) stationary waves. There may therefore be a succession of such clouds stretching downwind of the mountain. They often have very smooth outlines and are called *lenticular* (lens-shaped) or *wave* clouds (see pl. iv(*a*)). Since the vertical currents in them are usually quite strong they are greatly favoured by glider pilots.

Tropical clouds

Much of the energy for driving the global winds is supplied to the atmosphere by the tropical oceans from which vast quantities of moisture are evaporated and pumped to higher levels by the cloud systems, eventually to condense and provide energy, mainly in the form of latent heat, for the creation and maintenance of the great world-wide wind systems.

Over vast areas of the subtropical oceans, between 10° and 30° latitude, where the trade winds blow, cumulus clouds build by day and by night. Whereas, over land, cumulus clouds tend to form over selected breeding sites which have been preferentially heated by the sun, no such 'hot spots' exist over the open oceans. On the contrary, the sea-surface temperature is remarkably uniform. The trade-wind cumuli apparently come into being when random parcels of air, or eddies, normally swirling about in the subcloud layer, reach their condensation level, usually about 2000 ft (600 m) above the sea surface. Clusters of very small cloudlets, formed in this manner, somehow become organized into cumulus which then acquire many of the visual characteristics of similar-sized clouds forming over land.

A characteristic feature of the trade-wind cumulus is their tendency to form in lines or rows lying roughly along the direction of the wind, and to lean backwards, by as much as 45° to the vertical, in consequence of the wind strength decreasing quite rapidly with height in the cloud layer. This is well illustrated in pl. iv(*b*). Each cloud lasts for only about 20 min, individuals in the population are continually forming, growing and dissolving to be replaced by others. The rising volumes of air are limited and the clouds are strongly eroded by mixing and evaporating into the clear air.

They rarely rise above 7000–10000 ft (2–3 km) because, at these

Plate IV. (*a*) Lenticular wave clouds formed by the ascent of air over a mountain barrier. (Photograph by courtesy of Betsy Woodward.) (*b*) Cumulus clouds formed over the trade-wind oceans. The clouds lean backwards because the wind, which blows from left to right across the picture, decreases in strength at higher levels. (Photograph by courtesy of Dr Joanne Simpson.)

levels, they are checked by the presence of the trade-wind inversion*
which marks the beginning of a deep, stable layer of much drier air.
Some of these small, shallow clouds nevertheless produce showers,
but since the cloud tops are usually well below the level at which the
temperature falls to 0 °C (usually at about 15000 ft) they do not
develop the ice-crystal 'false cirrus' crowns which are so typical of
shower clouds outside the tropics. Although the inversion may be
imagined as a kind of lid hindering further growth of the clouds,
some of the larger and more vigorous nevertheless manage to penetrate it and supply moisture to the dry air above. This occurs specially
in the neighbourhood of the stormy disturbances which develop in
the trade-wind current, where the inversion is usually higher and
weaker than the average, so that the clouds can then build to considerable heights. They reach their full majesty during the development of the dreaded tropical cyclones (also called hurricanes in the
Atlantic and typhoons in the North Pacific) in which large cumulonimbus tend to be concentrated in a series of squall lines, each of which
spirals cyclonically about the rainless eye of the storm. The airman
flying into the eye finds himself in a huge natural amphitheatre,
surrounded by a great ring of thunderclouds 50 miles (80 km) or more
in diameter, the walls rising to heights of more than 40000 ft (12 km),
and the whole system surmounted by an enormous shield of cirrostratus and altostratus extending for hundreds of kilometres (see
pl. v).

The greatest thunderstorm activity, with giant clouds towering
up to 20 km above the ground and whose debris of anvil cirrus may
be stretched out in great sheets hundreds of kilometres long, is to
be found in the equatorial regions of the East Indies, the Congo
and Brazil. It is in this equatorial belt, between latitudes 10° N.
and 10° S., that the great influx of moisture from the trade-wind
region is pumped aloft and transported to higher latitudes, mainly by
the convective clouds.

Satellite pictures reveal that these clouds are often formed in giant
organized clusters as much as 1000 km in diameter which last for
a few days. They are composed of mesoscale units up to 100 km
across which, in turn, contain a number of cumulonimbus cells of
1–10 km diameter, but little is known about their internal structure,

---

* An inversion is a level at which the air no longer gets colder at greater heights;
on the contrary, the temperature increases with increasing height. Under these conditions the air will be stable for vertical motions in that a rising parcel of air will
become increasingly colder relative to its surroundings and therefore less buoyant.

Plate v. An organized system of cumulonimbus clouds formed in a great wall surrounding the central eye of a tropical cyclone (hurricane). The photograph was taken at a height of 5 km inside the eye, which is a region of light winds and little cloud, and gives a panoramic view of the steep sides of the encircling wall of the cumulonimbus, whose tops were at about 11 km. (Photograph by courtesy of R. H. Simpson and National Weather Records Center, North Carolina.)

organization and dynamics. Discovery of this, and how the formation of the clusters is related to the development of larger-scale disturbances in the equatorial wind field, were the main objectives of a very large, complex international expedition to the tropical eastern Atlantic in June–September 1974.

The causes of cloud formation

Clouds form primarily as the result of vertical motions in the atmosphere, the study of which is one of the most difficult and fundamental in meteorology. Clouds can be classified into four kinds according to the ascending motions which produce them:

   (i) Layer clouds formed by widespread regular ascent.

   (ii) Layer clouds caused by irregular stirring motions.

   (iii) Convective clouds.

   (iv) Clouds formed by orographic disturbances.

(i) Layer clouds formed by widespread regular ascent

The widespread layer clouds associated with cyclonic depressions, near fronts and in other bad-weather systems, are freqeuntly composed of several layers extending up to 30000 ft (9 km) or more, separated by clear zones which become filled in as the rain or snow develops. These clouds are formed by slow but prolonged ascent throughout a deep layer of air, a rise of a few centimetres per second being maintained for at least several hours. In the neighbourhood of fronts the vertical velocities become more pronounced and may reach about 10 cm/s.

Most of the high cirrus clouds visible from the ground lie on the fringes of cyclonic cloud systems and, though due primarily to regular ascent, their pattern is often determined by local wave disturbances which finally trigger their formation after the air has been brought near its saturation point by the large-scale lifting.

(ii) Layer clouds caused by irregular stirring motions

On a cloudless night, the ground radiates heat into space without heating the air adjacent to the ground, and thereby cools. If the air were very still, only a very thin layer would be chilled by contact with the ground. More usually, however, the lower layers of the air are stirred by motion over the rough ground, and the cooling is distributed through a much greater depth. Consequently, when the air is damp or the cooling is great, a *fog* several hundred feet deep may form, rather than a dew produced by condensation on the ground. Fog may also form when the air flows gently over progressively colder surfaces of land or sea; it is then called an advection fog in contradistinction to the radiation fog just described.

In moderate or strong winds the irregular stirring near the surface distributes the cooling upwards, and the fog may lift from the surface to become a *stratus* cloud which is not often more than 2000 ft (600 m) thick.

Radiational cooling from the upper surfaces of fogs and stratus clouds promotes an irregular convection within the cloud layer and causes the surfaces to have a waved or humped appearance. When the cloud layer is shallow, billows and clear spaces may develop so that it is described as stratocumulus instead of stratus.

(iii) Convective clouds

Usually cumuliform clouds appearing over land are formed by the rise of discrete masses of air from near the sun-warmed surface. These rising lumps of air, or *thermals* as they are called, may vary from a few tens to hundreds of metres in diameter as they ascend and mix with the cooler, drier air of their immediate surroundings.

Above the level of the cloud base the release of latent heat of condensation tends to increase the buoyancy of the rising masses which tower upwards and emerge at the top of the cloud with rounded upper surfaces as shown in pl. 11(*b*).

Since we cannot readily observe the detailed motions of thermals, either in clear air or inside the cloud, our ideas of what actually happens are based largely on model experiments (such as are described on page 19) and of our knowledge of fluid behaviour. Accordingly, we picture the development of a cloudy thermal as follows.

As the warmer, buoyant thermal pushes up through the cooler, denser air, the surfaces of constant air density become tilted relative to the horizontal surfaces of constant pressure and, in consequence, the air in the thermal is made to rotate and produce a vortex as shown in fig. 3. Continued condensation and release of latent heat will tend to increase the buoyancy and vertical velocity of the thermal. On the other hand, these tendencies are opposed by the mixing of the thermal with the surrounding air which occurs mainly across the advancing cap where small protuberances develop and engulf pockets of the relatively cool, dry environment. The thermal continually sheds its outer layers as fresh material emerges at its front surface and thereby warms and dampens the immediate surroundings. The thermal thus becomes more dilute, less buoyant, and therefore slower moving than would be the case if it did not mix with its surroundings.

At any moment a large cloud may contain a number of active thermals, and for the rest, be composed of the residues of earlier ones. A new thermal rising into a residual cloud will be partially protected from having to mix with the cool, dry environment and may therefore rise farther than its predecessor. Once a thermal emerges as a cloud turret at the summit or the flanks of the cloud, rapid evaporation of the droplets chills the cloud borders and destroys the buoyancy and even produces sinking. A cumulus therefore has a characteristic pyramidal shape, and viewed from a distance, appears to have an unfolding motion with fresh cloud masses continually emerging from the interior to form the summit, and then sinking aside and evaporating.

The depth of the clouds is determined by their initial buoyancy, the details of the mixing process and by the distribution of temperature and humidity in the surrounding air. In the lower parts of the layer occupied by these clouds the decrease of temperature with height in the clear air is usually more rapid than inside the clouds. Consequently the ascending volumes remain warmer, less dense and more buoyant than their immediate environment. Such a part of the atmosphere is said to be convectively unstable if it contains cloud, for the slightest upward movement of the cloud gives it buoyancy and it continues to rise. In the upper part of the layer the decrease of temperature upwards is usually less in the clear air than in the rising cloudy masses so that their excess temperature and buoyancy steadily diminish. The tops of the biggest clouds usually lie near the level where the excess buoyancy disappears altogether.

In settled weather, cumulus clouds are well scattered and small, with horizontal and vertical dimensions of only a kilometre or two. In disturbed weather, they cover a large part of the sky, and individual clouds tower up 10 km or more, often ceasing their growth only upon reaching the very stable stratosphere. These are the clouds which produce heavy showers, hail, and thunderstorms.

At the level of the cloud base the speed of the rising air masses is usually about 1 m/s, but may reach 5 m/s, and similar values are measured inside the smaller clouds. The up-currents in thunder-clouds, however, often exceed 5 m/s and may reach 30 m/s or more.

(iv) Clouds formed by orographic disturbances

These, the lenticular or *wave clouds* (see pl. IV (*a*)), are produced by the ascent of air over hills and mountains. The motion of the air may be disturbed at levels far above the mountain tops, and the vertical displacement may amount to a kilometre or more in favourable conditions. Where the air returns to its original level in the lee of the mountains, the clouds evaporate; forming in the crests of the wave-like disturbances set up by the mountain, they often strikingly demonstrate the nature of the air motion. Usually they are rather thin clouds, up to a kilometre thick, but when they occur in deep, cyclonic cloud systems, the wave motions intensify the condensation of moisture and cause the well known increase of rainfall over high ground. Thin wave clouds may occur at great heights (up to 10 km, even over hills a few hundred metres high), and are occasionally observed even in the stratosphere (at 20–30 km) over the mountains of Norway, Scotland, Iceland and Alaska. These stratospheric wave clouds are known as nacreous or 'mother-of-pearl' clouds because of the brilliant irridescent colours they display. Wave clouds are unlikely to appear at high levels unless the wind increases with height, and they are therefore usually associated with strong winds which carry the cloud particles through them rather quickly. Thus, although the cloud as a whole may remain almost stationary for several hours, the life of the individual cloud particles may be only a few minutes.

Imitating clouds in the laboratory

There are serious difficulties in attempting to simulate cloud formation in the laboratory largely because it is impossible to scale down all the physical quantities in the correct proportions and because of the disturbing influences of the walls of the containing apparatus. Nevertheless, some valuable clues on the manner of cloud formation and evolution may be obtained from model experiments even though they cannot reproduce the natural conditions in all respects.

(i) Simulation of altocumulus

We have seen that shallow layers of altocumulus cloud often break up into globules which may be arranged in rows whose alignment is definitely related to the wind shear at the cloud level. That these patterns are produced by convective overturning of the air in the cloud layer, caused by radiational cooling from its upper surface, is strongly suggested by the fact that similar patterns develop when a thin layer of gas is *heated* from *below*.

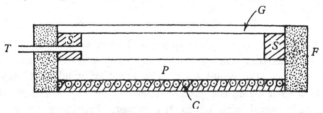

Fig. 1. Apparatus for producing convection cells which simulate, to some extent, shallow layers of convective cloud.

A convenient apparatus to demonstrate this may be easily constructed on the lines of fig. 1. *P* is a very smooth brass plate, about 30 cm dia., the temperature of whose upper surface may be kept constant at any value in the range 20–100 °C by varying the electrical current through the heating coil *C* wound on its undersurface. Above the plate is mounted a plane sheet of glass *G* supported on a brass-ring spacer *S*, the height of which can be varied between 2 and 16 mm. The assembly is now made into a fairly airtight chamber by sealing the joints with a little vacuum grease and lagging the sides with felt *F*. The air in the experimental space between the metal plate and the glass top is made visible by injecting cigarette smoke through the tube *T*.

No motion will be observed through the glass top unless the temperature difference between the top and bottom of the chamber (conveniently measured with thermocouples) exceeds a certain critical value which, for a chamber 10 mm deep, is about 12 °C. When this value is exceeded, the injected smoke first forms into long rolls and then, if conditions are very steady, soon breaks up into a series of polygonal convection cells having usually five or six sides as shown in fig. 2 and pl. VI. The motion of the air is downwards in the centre of the cells, upwards at the outer margins, and inward at the top. If the chamber is less than 7 mm deep, the polygonal cells form almost instantaneously, but if it is less than 6 mm, no polygonal cells form whatever the temperature régime.

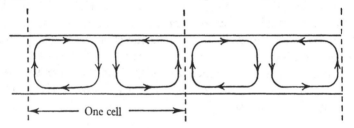

Fig. 2. The circulation in convection cells.

This experiment simulates quite well certain features of alto-cumulus formation when the wind is steady and varies little with height through the cloud layer. The effect of windshear may be simulated by drawing the glass plate across the top of the chamber at a steady speed with an electric motor. Providing the critical temperature gradient is exceeded, the smoke now arranges itself in a series of rolls with adjacent rolls rotating in opposite directions. In strong shear (with the plate moving rapidly), the rolls are arranged along the direction of shear; when the shear is weak the rolls are aligned at right angles to the shear and are therefore reminiscent of billow clouds. This experiment is both instructive and suggestive for studying convection in a shallow layer of gas. However, because viscosity and molecular heat transfer are much more important in this small-scale experiment than in the atmosphere, it does not provide an exact analogue for the development of altocumulus.

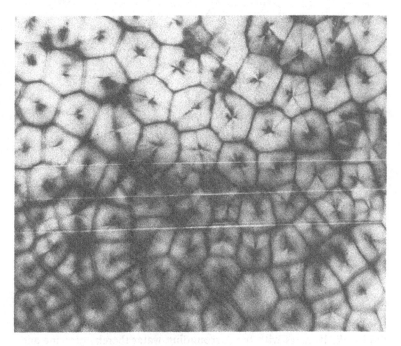

Plate VI. Polygonal convection cells produced in a thin layer
of smoke-filled air heated from below.

The apparatus is fairly simple and cheap to make, but it requires
a little manipulative skill to obtain good results.

(ii) A model of cumulus

A very life-like model of a cumulus cloud may be produced by
suddenly releasing a blob of a white and slightly denser liquid
representing the cloud into a large tank of still, clear water representing the clear atmosphere. A rectangular glass tank (the larger
the better, but one $12 \times 12 \times 24$ in. deep and open at the top will be
suitable) is needed for the experiment. The cloud is created by suddenly overturning a small hemispherical cup pivoted about a horizontal axis through its centre and which is partly immersed in the
surface of the water and filled with the cloudy liquid up to the level
of the surface of the tank. The cup may be made by cutting in half a
hollow plastic ball and a suitable cloudy liquid is salt solution made
visible by a dense white precipitate produced on mixing solutions
of barium chloride and sodium sulphate. The initial density difference

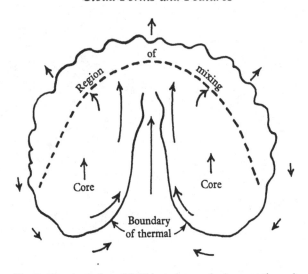

Fig. 3. The circulation of fluid in and around a buoyant thermal
developing as a vortex.

between the 'cloud' and the environment can be controlled by
selecting the concentration of the solution. As the blob of denser
fluid sinks it mixes with the surrounding water thereby growing and
becoming more dilute. Protuberances develop on the advancing
cap due to erosion by the surroundings and soon the falling mass
takes on much of the visual character of a cumulus cloud, but growing
up-side down (see pl. VII). Mixing with the environment occurs
predominantly at the cap, the mixed material flowing round to be
incorporated into the rear of the bubble, which now develops as a
vortex as shown in fig. 3. Certainly the visual appearance and
development of this artificial cloud are similar in many respects to
those of a growing cumulus, and such experiments have been useful
in helping us to visualize the air motions which may occur inside
a natural cloud. But because the density differences are rather
exaggerated in the model experiment, and because it is very difficult
to simulate such factors as an unstable atmosphere, wind shear, and
the sudden release of latent heat of condensation, the analogy must
not be pressed too far.

Cloud photography

Photographing clouds can be a very instructive and rewarding
hobby. It is not easy to generalize about what makes a good cloud

Plate VII. A model cumulus produced by releasing a blob of dense white liquid into a tank of still, clear water. (Photograph by R. S. Scorer.)

picture, but the photographs in this book indicate what may be achieved by an observer with an eye for the significant and interesting features of a cloudscape. In general, the clouds should dominate and fill most of the picture. Other objects may be included to give direction and perspective but these should occupy the margins of the picture. In order to recognize, select and describe suitable subjects for photography, it is necessary to understand something of cloud evolution and of the physical processes involved. In this regard, the reader cannot fail to profit from a little book called *Cloud Study** which contains a selection of excellent pictures, admirably described and explained. A beautiful but much more expensive collection of photographs, mostly in colour, and accompanied by excellent interpretations and explanations, appears under the title *Clouds of the World*† by R. S. Scorer.

Still pictures and movies of clouds in black-and-white are best taken with a fine-grain film such as FP 4 and a red filter for improved contrast. When photographing distant clouds in directions well away from the sun both definition and contrast may be improved by

* *Cloud Study*, by F. H. Ludlam and R. S. Scorer. (London: Murray. Price £1·75.)

† *Clouds of the World*, by R. S. Scorer. (Newton Abbot: David and Charles. Price £12.60.)

rotating a polaroid in front of the camera lens to reduce the intensity of the light scattered from haze particles in the atmosphere.

The growth and development of clouds can be best studied from speeded-up time-lapse pictures. The evolution of a growing cumulus may be photographed, on cine-film, say at the rate of 1 frame every 2 s; if the film is then projected at 16 frames/s, one observes the cloud motions speeded up 32 times.

The quality and appeal of cloud photographs may be greatly enhanced by the use of colour. With colour film, no red filter is required, but exposure times are more critical than for black-and-white and an exposure meter is almost essential. Still pictures in colour are usually prepared as 35 mm transparencies. Time-lapse pictures may be taken on either 8 mm or 16 mm film.

When photographing halos and other optical phenomena near the sun, it is usually necessary to obscure the sun either by making use of cloud, obstacles such as telegraph poles, or by mounting a small circular black disk on an arm in front of the lens.

2

THE NUCLEI OF CLOUDY
CONDENSATION

Introduction

Clouds are formed by the lifting of damp air which then cools by
expansion as it encounters continuously falling pressures at higher
levels. The relative humidity* thereby increases and, if the air were
entirely free of foreign particles and of electrically charged clusters
of molecules, called ions, this would continue until the pressure of
the water vapour became about eight times that required to *saturate*†
the air. In this highly (sevenfold) *supersaturated*‡ air, water droplets
form *spontaneously* as water-vapour molecules become arranged in
clusters as the result of microscopic random fluctuations in the
density and temperature of the vapour. This phenomenon was
discovered by C. T. R. Wilson in 1897 when he rapidly expanded
clean, particle-free, water-saturated air in the cloud chamber which
was later to play such an important role in the detection of atomic
particles.

Atmospheric clouds are not formed in this manner because the
air is never completely clean; it usually contains a wide variety of
airborne particles (aerosols) on which the water vapour may condense when the air is only very slightly supersaturated or even slightly
undersaturated. The first experiments to demonstrate condensation
of water vapour on airborne particles were carried out in France by
Coulier in 1875 and by John Aitken working in Falkirk in 1880.
Aitken's researches, which were conducted both in the laboratory
and in the open air between 1880 and 1916, were of the greatest
importance. His 'dust' counter, in which the sudden expansion of a
known volume of air caused droplets to form on some of the

* The relative humidity, H, or the saturation ratio, S, of the air, is defined as the
ratio of its actual vapour pressure, p_V, to that required to saturate the air at the same
temperature, p_s, Thus H or $S = p_V/p_s$, and when the air achieves saturation, $p_V = p_s$
and $S = 1$. The relative humidity is commonly expressed as a percentage.

† Air is saturated when there is no *net* transfer of vapour molecules between it and
a plane surface of water at the same temperature.

‡ The supersaturation σ of the air is given by $p_V/p_s - 1$; this may also be expressed
as a percentage by multiplying by 100. Thus air which has a saturation ratio of 1·01
(corresponding to a relative humidity of 101 %) has a supersaturation of 0·01 or 1 %.

particles, was really a prototype of the Wilson cloud chamber. The droplets settled on to a stage and, when counted under a low-power microscope, gave the number of particles in unit volume of air which acted as centres of condensation or, as we now say, *condensation nuclei*. With this simple apparatus, Aitken discovered that there were two broad classes of nuclei; those having an affinity for water vapour, i.e. *hygroscopic* particles, on which condensation begins even before the air becomes saturated, and *non-hygroscopic* particles which require some degree of supersaturation depending upon their size, to act as centres of condensation.

The degree of supersaturation achieved in a cooling cloudy air mass will depend upon the temperature and the rate of cooling, which control the rate at which vapour becomes available for condensation, and upon the concentration, size and nature of the aerosol particles which will govern the rate at which the vapour condenses.

The condensation nuclei therefore play an essential role at the very beginning of the cloud-forming process; let us now inquire into their origin and physical behaviour.

The behaviour of a condensation nucleus

The equilibrium vapour pressure p_r over the surface of a droplet of pure water of radius r exceeds that, p_∞, over a plane water surface at the same temperature according to the equation

$$\log_e p_r / p_\infty = 2\gamma M / \rho_L R T r,* \qquad (2\cdot1)$$

where γ, ρ_L, M are respectively the surface tension, density and molecular weight of water, R the universal gas constant and T the absolute temperature. Thus, if such a droplet is to remain in equilibrium with its surroundings, that is it neither evaporates nor grows, the vapour pressure of the surrounding air must equal p_r of (2·1), and will therefore be supersaturated. The degree of supersaturation required will be higher the smaller the droplet, some typical values, calculated from (2·1), being given in table 2 (*b*).

Thus a droplet of radius one ten-millionth of a centimetre

---

* Derivation of this expression for the equilibrium vapour pressure over a curved liquid surface, originally due to Lord Kelvin, may be found in many physics textbooks, e.g. Newman and Searle's *General Properties of Matter*, fourth edition, p. 198. (Edward Arnold and Co.). A more rigorous derivation is given in *The Physics of Clouds*, second edition, by B. J. Mason, p. 2. (Oxford: Clarendon Press, 1971.)

TABLE 2. *Critical radii and supersaturations for nuclei of various sizes at 273 °K*

(a) Hygroscopic nuclei of NaCl

log m (g) …	-16	-15	-14	-13	-12	-11	-10	-9	-8
$r(\mu)$ at $H = 78\%$	0·039	0·084	0·185	0·39	0·88	1·85	4·1	8·8	18·5
$r_c(\mu)$*	0·20	0·62	2·0	6·2	20	62	200	620	2000
$H_c - 100$ (= supersat. %)†	0·42	0·13	$4·2 \times 10^{-2}$	$1·3 \times 10^{-2}$	$4·2 \times 10^{-3}$	$1·3 \times 10^{-3}$	$4·2 \times 10^{-4}$	$1·3 \times 10^{-4}$	$4·2 \times 10^{-5}$
r (of crystal) (μ)	0·022	0·048	0·103	0·22	0·48	1·03	2·2	4·8	10·3

r at $H = 100\%$ is approx. $r_c/\sqrt{3}$.
* For other nuclear substances of molecular weight M_1, multiply by $(58·5/M_1)^{\frac{1}{3}}$.
† For other nuclear substances of molecular weight M_1, multiply by $(M_1/58·5)^{\frac{1}{3}}$.

(b) Pure water droplets or insoluble wettable nuclei

log r (cm) …	-7	-6	-5	-4	-3
$100\, p_r/p_\infty$	323	112·5	101·2	100·12	100·01

(10^{-7} cm) requires a saturation ratio of 3·23 or a supersaturation of 223 % to persist, while droplets of radius greater than 10^{-5} cm require supersaturations of less than 1 %.

Table 2 shows the supersaturations which must be exceeded for continued condensation to occur on pure water droplets; for insoluble, wettable particles of the same size, the supersaturations required will be slightly less, while those required for the activation of water-repellant particles will be slightly higher.

If the droplet is formed on a wholly or partially soluble nucleus, the equilibrium vapour pressure at its surface is reduced by an amount depending on the nature and concentration of the solute, which means that condensation is able to set in at lower supersaturations than those required for an insoluble particle of the same size. This is a fact of considerable importance in cloud physics because many airborne particles are composed wholly or partly of soluble matter.

The equilibrium vapour pressure p'_r over a droplet of pure solution of radius r is given, very nearly, by the equation*

$$p'_r/p_\infty = \exp(2\gamma M/\rho_L RTr)\left[1 - \frac{8 \cdot 6m}{M_1 r^3}\right], \qquad (2 \cdot 2)$$

where m is the mass of the dissolved salt in grammes, M_1 the molecular weight of the salt, the other symbols being defined as in (2·1).

Equation (2·2) can be used to calculate the relative humidity H, which the air must have to remain in equilibrium with nucleus droplets of a given radius containing a specified mass of a particular salt or, conversely, to compute the radius of drops which will be in equilibrium with an atmosphere of a given relative humidity. We have, then, $p'_r/p_\infty = H/100$, and if the humidity is nearly 100 %, (2·2) may be written more simply as

$$\frac{p'_r}{p_\infty} = 1 + \frac{2\gamma M}{\rho_L RTr} - \frac{8 \cdot 6m}{M_1 r^3}. \qquad (2 \cdot 3)$$

Graphs showing how the equilibrium radii of droplets containing specified masses of sodium chloride vary with the relative humidity are shown in fig. 4. Some growth of these hygroscopic particles occurs before the air becomes saturated, and indeed they become droplets of saturated sodium chloride solution at 78 % relative

---

* For the derivation of this equation, originally due to H. Köhler of Sweden, see *The Physics of Clouds*, p. 24.

humidity. The equilibrium radius of the droplet increases with increasing humidity until the air becomes supersaturated by a critical amount ($H_c - 100$) corresponding to the maximum of the curve in fig. 4. At this stage the solution is quite dilute, and if the droplet exceeds this critical radius given by

$$r_c = \sqrt{(8 \times 10^5 mT/M_1)},$$

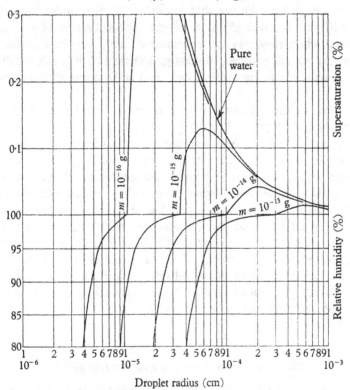

Fig. 4. The equilibrium relative humidity (or supersaturation) as a function of droplet radius for solution droplets containing the indicated masses of sodium chloride.

the supersaturation required to maintain equilibrium thereafter *decreases* with increasing droplet size. If the supersaturation were maintained, transition from a *nucleus droplet* of radius less than r_c to a fully developed *cloud droplet* of radius greater than r_c would now occur very rapidly and, in theory, the droplet would grow without limit. In practice, we are not concerned with a droplet growing in isolation under a steady supersaturation, but rather with a whole population of droplets competing for the water vapour being released in a cooling air mass. In this case, the supersaturation will not, in general, remain steady; when the vapour is being extracted at a faster rate than it is being released, the supersaturation is forced to retreat and so growth of the droplets is restricted.

Table 2 shows the critical radii and critical supersaturations calculated from (2·2) for sodium chloride nuclei with masses between 10^{-16} and 10^{-8} g. We see that nuclei of 10^{-15} g achieve radii of $6 \cdot 2 \times 10^{-5}$ cm (or $0 \cdot 62 \, \mu^*$) at a critical supersaturation of $0 \cdot 13 \%$ beyond which they may act as centres of continued condensation. The very big salt particles, of mass greater than 10^{-11} g, will usually not remain in the atmosphere long enough to attain their critical size.

So far, we have discussed the behaviour of completely insoluble particles and of purely hygroscopic, soluble particles as potential condensation nuclei. But atmospheric aerosols are often partly insoluble; over the continents we often find insoluble particles coated with a thin layer of hygroscopic substance. At high humidities, these *mixed nuclei* react to humidity changes like wholly soluble particles of equivalent size, but at humidities below about 70 %, the solution coat shrinks until the particle becomes almost completely solid. The entirely soluble and the mixed nuclei, being hygroscopic, will obviously be preferred as centres of cloud-droplet formation. Whether or not non-hygroscopic particles will play a significant role in cloud formation will depend largely on whether hygroscopic nuclei are present in sufficient numbers to absorb the available water vapour.

The abundance and sizes of condensation nuclei

Atmospheric aerosols cover a wide range of particle sizes, from about 10^{-7} cm radius for the small ions consisting of a few neutral air molecules clustered around a charged molecule, to more than 10μ (10^{-3} cm) for the largest salt and dust particles. Their concentrations, expressed as the number of particles per cubic centimetre of air, also cover an enormous range. The small, ubiquitous ions almost certainly play no role in atmospheric condensation because of the very high supersaturations (about 300 %) required for their activation, while the largest particles are able to remain airborne for only a limited time.

* $1\mu = 1$ micron $= 10^{-4}$ cm. This unit will be used frequently from now on.

Because of the wide variety of particle size and concentration, several different experimental techniques are necessary to cover the whole range of atmospheric aerosols. For this reason, and because the collection, sizing and examination of the particles is extremely laborious, measurements have been made at only a relatively few

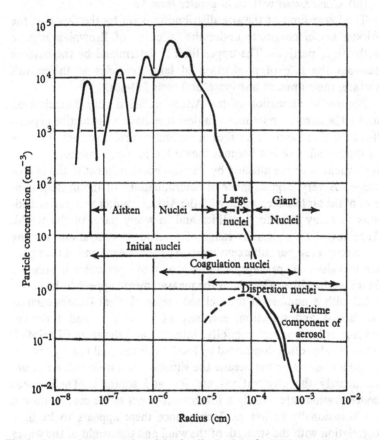

Fig. 5. A general representation of the size distribution of natural aerosols in heavily polluted air over land. (From Junge, *Berichte der Deutsch Wetterdienst. U.S. Zone,* **35** (1952), 261.)

places. The data are not therefore representative of the whole atmosphere under a variety of meteorological conditions but, as far as they go, they suggest that the size distribution of aerosols measured over land is, on average, like fig. 5. It is convenient to divide the particles into three size groups.

(i) Nuclei having radii between 5×10^{-7} and 2×10^{-5} cm $(0\cdot2\mu)$ which are called *Aitken nuclei* in recognition of the fact that particles of this size, but not the very small ions, are detected in the Aitken 'dust' counter.

(ii) *Large nuclei* with radii between $0\cdot2$ and 1μ.

(iii) *Giant nuclei* with radii greater than 1μ.

The lower limit of the size distribution is set by the fact that the Aitken nuclei coagulate under the influence of Brownian motion with larger particles. The upper limit is determined by the balance between the rate of production of large particles at the earth's surface, their upward transport, and their fall-out.

Positive identification of the Aitken nuclei is very difficult even under the electron microscope unless they have a distinctive crystalline form like the perfect cubes of sodium chloride, not only because of their small size but because they often partially evaporate in the high vacuum or are affected by electron bombardment in the microscope. A very large number of determinations of the nucleus content of the air has been made with the Aitken counter and its descendants in many different localities and in many parts of the world. They reveal an enormous range of particle concentrations varying from only a few per cubic centimetre over the oceans and in the upper air to values of, perhaps, a million or more per cubic centimetre in industrial cities. Over land the measurements appear to be correlated with a number of interrelated meteorological factors such as windspeed and direction, intensity of convective and turbulent mixing, solar heating, humidity, intensity and duration of rainfall. They are also often dominated by local influences and nearby sources of pollution. Over the oceans the situation is less complex but unfortunately the observations are few and sporadic. The nucleus concentrations are usually a few hundreds per cubic centimetre but are occasionally as low as $2/cm^3$; since there appears to be little correlation with the strength of the wind and the height of the waves it seems unlikely that they originate from the sea surface.

A number of methods have been devised for collecting the large and giant nuclei for subsequent examination under the optical or the electron microscope. One may employ an air jet of high velocity, suction through special filters, or precipitation by electrical or thermal forces to ensure efficient collection of particles on a relatively large surface, or alternatively, use very narrow collectors (e.g. spiders' threads) and correspondingly low air velocities.

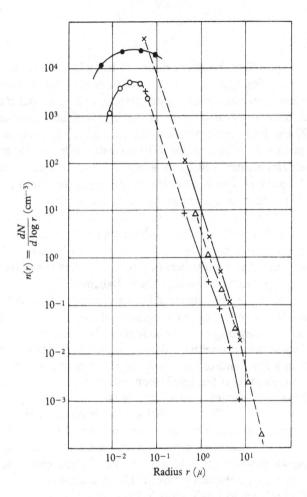

Fig. 6. The size distribution of continental aerosols at various sites in Germany. (After Junge, *Tellus*, **5** (1953), 1.)

The average abundance of large and giant particles of continental aerosol, expressed as the number of particles found in a cubic centimetre of air, and their size distribution are shown in fig. 6. The size distribution shows a certain regularity which is a result of dynamical equilibrium between production of particles from a variety of sources and their loss by coagulation, sedimentation, capture and washout, and follows the law

$$n(r) = \frac{dN}{d(\log r)} = \frac{A}{r^3}, \tag{2·4}$$

or
$$N' = A'/r^3 = B/m, \tag{2·5}$$

where $dN = n(r)d(\log r)$ is the number of particles per cm³ in the radius interval $d(\log r)$, N' is the total number of particles per cm³ of radius greater than r and mass greater than m, and A, A' and B are numerical factors. Measurements made in Frankfurt, Germany, about 400 km from the sea, show the number concentration of particles greater than $0·1\ \mu$ radius to be typically $10^9/m^3$ or $10^3/cm^3$, that of particles greater than 1μ to be about $10^6/m^3$ or $1/cm^3$, while particles larger than 10μ are present to the extent of only $10^3/m^3$ or $1/litre$*. The large nuclei contribute, on average, about 100 particles/cm³, and the giant nuclei about $1/cm^3$ compared with Aitken-nucleus counts of about $40000/cm^3$. Measurements in open country, for example, on the summit of the Puy de Dôme in France, and in Greenland, show that the numbers of large nuclei vary considerably from day to day, but again average about $100/cm^3$.

Over the oceans, the abundance of large and giant nuclei increases with increasing wind speed and roughness of the sea surface, for as we shall see, they originate in the foam of breaking waves. The results of some measurements made on board ship and from aeroplanes flying a few hundred feet above the sea surface are shown in fig. 7 which suggests that the distribution reaches a peak at a mass of about 10^{-14} g corresponding to a dry crystal radius of $0·1\ \mu$, and ranges from about 2×10^{-15} g to 10^{-8} g. Over a calm sea the total concentration of salt particles rarely exceeds 1 cm⁻³ or 1% of the total numbers of condensation nuclei active in a cloud chamber at $0·5\%$ supersaturation. Even over a rough sea, the concentration of salt nuclei of $m > 10^{-11}$ g rarely exceeds $1/cm^3$, and the *total* concentration of all salt nuclei rarely exceeds $10/cm^3$.

The aerosol is likely to be entirely of maritime origin only when the air has a long, uninterrupted track over the ocean; more often it will be contaminated with some particles of land origin and so have a mixed or transitional character. High concentrations of salt nuclei, produced by surf, occur near the coastlines, but their numbers fall off quite rapidly during the first few miles inland and thereafter more slowly as the particles become distributed vertically by turbulent mixing in the lower layers of the atmosphere. In the absence of

---

* The customary abbreviation '1.' for litre is used henceforth, e.g. 1/1. not 1/litre.

rain, which washes out the larger particles, appreciable numbers of salt nuclei may be found some hundreds of miles inland at heights of a few hundred feet although, near the ground, many may be lost as they fall out or are captured by trees and similar obstacles.

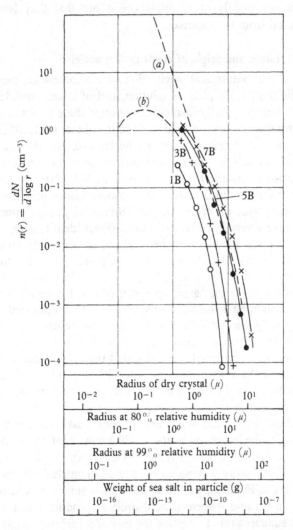

Fig. 7. Average size distribution of sea-salt nuclei measured over the oceans for wind forces 1, 3, 5, 7 on the Beaufort scale according to A. H. Woodcock. The line (*a*) gives the size distribution of continental aerosol for comparison. The line (*b*) is an extrapolated size distribution of marine aerosol. (From Junge, *Atmospheric Chemistry, Adv. Geophys.* **4** (1958), 1.)

The Aitken nuclei, which have a mean life in the atmosphere of several days, may be carried large distances from their source. The fact that Aitken-nucleus counts over the ocean do not change appreciably with wind speed and the state of the sea surface indicates that they are not mainly of maritime origin but that they have been transported from land sources.

Nature and origin of condensation nuclei

Turning now to the nature and origin of condensation nuclei, they consist of solid matter, droplets of solution, and of mixed particles that are partly soluble, partly insoluble. Although there is virtually no *direct* information on the chemical constitution of the small Aitken nuclei, they are almost certainly formed by the condensation and sublimation of vapours in natural and man-made smokes and by chemical reactions between trace gases such as SO_2, NH_3 or Cl_2 produced either from natural sources or by combustion; since initial production from a wide variety of sources is followed by coagulation, the particles have a very mixed constitution. Direct identification of larger particles is possible by microchemical and electron-diffraction methods but is tedious. Thus Junge found that in industrial cities the 'large' soluble nuclei of $0.1\,\mu < r < 1\,\mu$ consisted largely of ammonium sulphate, while the 'giant' particles of $r > 1\,\mu$ were either mixed nuclei containing a good deal of sulphuric acid or consisted of chlorides, probably sea-salt. The origin of the ammonium sulphate and sulphuric acid particles remained obscure, but in the last ten years it has become increasingly probable that they are produced in solution following the absorption of gaseous SO_2 and NH_3 in cloud and fog droplets or by their adsorption together with water vapour on the surfaces of large solid particles.

The importance of sea-salt as a source of condensation nuclei has been disputed for the last 50 years and is still a matter for argument. Aitken was the first to suggest that the salt-dust resulting from the evaporation of sea-spray was an active source of nuclei, but such leading workers as Findeisen and Simpson argued against this being a major source because it seemed impossible to produce sea-spray droplets at a sufficient rate to replace the loss of nuclei in precipitation. It is true that the spray drops torn off the crests of breaking waves are generally too large to remain airborne for long, but in recent years Woodcock and his co-workers have demonstrated by high-speed photography that salt particles of $r > 1\,\mu$ are produced mainly by the bursting of numerous small air bubbles in the foam of breaking waves. According to them, the smallest bubbles, of $R \sim 50\,\mu$, produce salt particles of $m \sim 10^{-11}$ g; larger bubbles produce larger particles of mass up to 10^{-5} g. The particles are formed by the evaporation of drops that result from the break-up of small liquid jets that rise from the cavities left by bursting bubbles—see fig. 8.

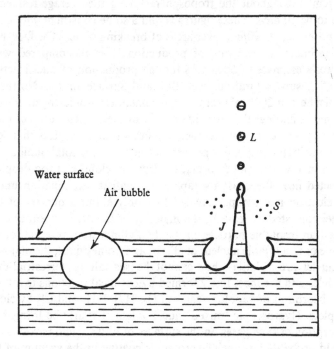

Fig. 8. The formation of sea-salt particles from the bursting of bubbles. The large drops L are formed by disintegration of the jet J. More numerous smaller particles S are formed by bursting of the bubble film.

But, in addition to these drops, Mason discovered that the disruption of the bubble caps produces much larger numbers of droplets too small to be detected except in the supersaturated atmosphere of a cloud chamber. This, and later work by Day, established that the average number of film droplets is dependent on bubble size, ranging from none for bubbles of $R < 50\,\mu$ to 300–400 for bubbles of $R = 2$ mm, the average number being $\sim 10^4 R^2$ where R is in cm. Combining this result with Blanchard and Woodcock's observations that the number of bubbles of radius between R and $R + dR$ bursting per cm^2/s in a foam patch is $3 \times 10^{-6} dR/R^4$, we arrive at a total rate of production of film droplets by bubbles of $R < 50\,\mu$ of $\sim 10\ cm^2/s$, and a similar estimate for the production of droplets by breaking bubble jets.

A production rate of $20/cm^2$ s in areas of breaking waves would imply an overall rate of about $1/cm^2$ s for the ocean as a whole and this would be consistent with an average salt-particle concentration of $1/cm^3$ throughout the troposphere taking the average residence time to be 10 days. These figures would also be consistent with concentrations of 10–$20/cm^3$ over areas of breaking waves. The fact that this estimate for the rate of production is small compared with Squires's estimate of $500/cm^2$ s for the production of nuclei active at 0.5% supersaturation over the land surface of the Northern Hemisphere or $200/cm^2$ s for the Hemisphere as a whole, together with Twomey's findings that even in oceanic air the concentration of non-volatile particles rarely exceeded $10/cm^3$, strongly suggests that sea-salt contributes only a few per cent, at most, to the total numbers of nuclei involved in cloud formation. Even in cloud and fog droplets collected near the coast by Japanese scientists, examination under the electron microscope revealed that only about a quarter of the nuclei consisted of sea-salt, the majority of the droplets either contained an insoluble particle, probably a combustion product, or no visible or identifiable nucleus at all. If, as Twomey finds, the majority of nuclei are volatile, they cannot be sea-salt or terrestrial-dust particles and are most likely formed by chemical reactions involving the absorption of vapours such as SO_2, H_2S, NH_3 and Cl_2 by cloud droplets. Hygroscopic 'mixed' nuclei might similarly be formed by adsorption of such gases on the surfaces of wetted solid particles but would probably lose their hygroscopic coating in the vacuum of the electron microscope.

Despite the increased scale of atmospheric pollution in modern times, one would expect natural sources of nucleogenic material to outweigh man-made sources unless the constitution of clouds has changed markedly over the last 200 years. Squires's estimate, based on measurements made in the United States, is that man-made sources in the Northern Hemisphere contribute only about 5% of the nuclei active at 0.5% supersaturation, the implication being that the major sources are natural combustion which release large quantities of SO_2 and particles into the air, the release of H_2S and NH_3 by decaying organic matter, and the escape of Cl_2 gas from the sea.

However a good deal more careful work is required to establish firmly the nature and origin of that part of the atmospheric aerosol that becomes involved in cloud formation and, in particular, the relative contributions made by land and sea, by natural and man-made processes.

Experiments

1. Condensation on atmospheric nuclei

Condensation on atmospheric nuclei, following expansion of the air, may be simply demonstrated as follows. A little water is placed in a glass flask fitted with an outlet tube and a tap, shaken, and left for several minutes with the tap closed to allow the air to become saturated. The outlet tube is now connected with rubber tubing to a bicycle pump, a few strokes of which raise the air pressure inside the flask. With the tap closed and the pump disconnected, the flask is again left for a few minutes to allow the air to achieve equilibrium. A narrow bright beam of light from a strong torch passes through the flask and if now one looks into the beam and opens the taps, a dense fog is seen. This fog consists of tiny droplets of water produced by condensation of vapour on nuclei present in the air which expanded to atmospheric pressure through the open tap.

Looking in the direction of the lamp, one may see coloured rings forming a corona—these being due to diffraction of the light by the droplets. A much thicker fog, consisting of much greater numbers of smaller droplets which produce brighter corona rings of larger diameter, may be made by introducing cigarette smoke into the flask. This provides a greater supply of condensation nuclei.

2. Measurement of the activity of condensation nuclei using a diffusion cloud chamber

Condensation nuclei can be detected, and the numbers which become activated at different levels of supersaturation can be determined, by using a simple chemical diffusion cloud chamber. The bottom of a small glass chamber is covered with hydrochloric acid of known strength and the top is closed by a porous plate or disc of filter paper soaked in distilled water; the whole chamber being kept at constant room temperature. Water vapour diffuses downwards and HCl gas diffuses upwards until a steady state is reached at which the intermediate air space becomes supersaturated with respect to the pure water surface. The maximum supersaturation achieved depends only on the concentration of the acid solution and values ranging from 0·1 to 100% can be produced and determined. In a chamber maintained at 20 °C, HCl solutions with concentrations of 1, 2·5 and 10% produce maximum supersaturations of 0·5, 1·0 and 10% respectively.

The numbers of nuclei that become activated at different supersaturations can be determined by introducing equal samples of the air under test into each of several beakers each containing acid of a different concentration.

In practice, any nuclei present in the chambers before the experiment are allowed to grow into droplets and fall out. When the air space is seen to be clear and free of droplets, a known volume of the nucleus-laden air under test is introduced into the top of a chamber through a valve, and the resulting droplets are viewed, counted or photographed as they fall through a well-defined and collimated beam of light from a strong source such as a mercury arc. In this way it is possible to determine the number of particles in, say, 1 cm³ of the sample that act as condensation nuclei at supersaturations of 0·5, 1·0, ..., 5%.

For further details see S. Twomey, *Geofisica pura e applicata*, **43** (1959), 227.

3. The growth of individual condensation nuclei with increasing humidity

The equilibrium radii of solution droplets, formed on nuclei greater than about 10^{-13} g, may be measured in a simple and elegant manner first used by the French cloud physicist, H. Dessens.

Salt crystals, collected from the atmosphere or from a fine spray of salt solution, are caught on fine spider's threads and placed in an airtight box in which the humidity of the air may be maintained at a constant value by bringing it into equilibrium with a salt solution of a given concentration. Windows in the top and bottom of the box allow the droplets formed around the salt crystals to be illuminated and viewed through a microscope. When their diameters show no further increase they are measured against a scale in the eyepiece. The humidity in the box may be changed by varying the concentration of the salt solution, or by using a different salt. Suitable saturated solutions for use at 20 °C are: KNO_3 (humidity 45 %), $NaNO_2$ (66 %), NH_4Cl (80 %), KBr (84 %), $ZnSO_4.7H_2O$ (90 %) and $Pb(NO_3)_2$ (98 %). In this way, the equilibrium radii of the droplets at various humidities may be found, and for salts other than sodium chloride. Although a sodium chloride solution becomes saturated when in equilibrium with air of 78 % relative humidity, small droplets may remain as *supersaturated* solutions at humidities as low as 40 %. Such supersaturated droplets may crystallize violently to produce one or several crystals.

A suitable collector may be made by encouraging a spider (a small garden variety is recommended) to spin a network of threads across a small, rectangular wire frame. A good experimenter will obtain a grid of parallel, equally spaced fibres about 0.01μ dia. This may be rotated, at constant speed, through the air or a cloud of salt spray by mounting it near the edge of a gramophone turntable. The air in the humidity box should be continually stirred with a small battery-driven fan to ensure uniform, equilibrium conditions.

Microscopic examination of the fibres will reveal the variety of size, shape, abundance and constitution of the larger ($r > 0.5\mu$) atmospheric particles, and the technique may be used to study the variations which occur, from day to day and from place to place, in the aerosol content of the atmosphere. The relative numbers of hygroscopic and non-hygroscopic particles may be determined by noting the fraction which grow under relative humidities of less than 100 %.

3

THE GROWTH OF CLOUD DROPLETS

Introduction

So far we have discussed the nature and properties of the condensation nuclei and the early stages of growth of an individual nucleus. We have now to consider the further growth of populations of nuclei leading to the formation of cloud or fog. The growth rate of individual droplets depends not only on the hygroscopic and surface tension forces and the humidity of the surrounding air, as already discussed, but also the rates of transfer of water vapour to and heat of condensation from the droplet. When we have to deal with a population of droplets the problem becomes more complicated; since all the droplets compete for the available water vapour, their growth rate will depend upon the concentration, size and nature of the nuclei, the rate of cooling of the air which controls its temperature and supersaturation, and the scale and intensity of the turbulent motions in the cloud. The influence of these various factors on the growth of cloud particles has been investigated from the theoretical point of view, as we shall see; but first, let us look at the measurements of droplet size and concentration of liquid water in clouds.

The sizes of cloud droplets

It may seem surprising that determination of the concentration and the sizes of cloud droplets, which are large enough to be easily seen under the microscope, should present a difficult experimental problem. But the truth is that, despite a good deal of thought and effort, we still do not have an entirely satisfactory technique for use from a moving platform such as an aeroplane.

In the more direct methods the droplets are caught on a suitably prepared and exposed surface and the diameter of the droplets or their impressions are later measured under the microscope. This method is extremely tedious since a collector, only 1 mm^2 in area, would collect several hundred droplets after only $\frac{1}{100}$ s exposure from an aircraft travelling at 250 m.p.h.; it is, however, the most reliable. The accuracy of the method depends upon obtaining a truly representative sample from the cloud, preventing evaporation of the droplets between sampling and counting, and avoiding shattering or coalescence of the captured droplets. In order to obtain a true sample, the collector should collect droplets of all sizes with equal efficiency but because, on approaching the collector, the droplets are deflected from their original paths to an extent depending upon the radii of the droplets, the airspeed, and the width of the collector, the latter tends to discriminate against the smaller droplets which are deflected around it.

In one much-used method, a small glass slide coated with oil is exposed to the cloud from an aircraft for about $\frac{1}{100}$ s by a spring-loaded gun. The impinging droplets penetrate into the oil (if this is of the right thickness and viscosity) and become completely submerged. Even so, the smallest droplets, only a few microns in diameter, tend to evaporate into the oil, making it necessary to photograph the slide under the microscope in the aircraft and take the measurements later from the photograph. A typical sample of cloud droplets recorded in this manner is shown in pl. VIII.

In another method, the droplets impinge on a slide coated with a thin, white layer of magnesium oxide in which they leave craters about one-third larger in diameter than themselves*. Some workers prefer soot-covered slides.

The direct photography of droplets at some distance from the aircraft skin poses conflicting requirements of a large magnification, sharply focused images, and a reasonable sampling volume (depth of field of view). Some of these problems are overcome in an imaging device developed by Knollenberg that measures the sizes of shadows cast on a linear array of optical fibres as the droplets pass through a collimated, high-intensity light beam. The array of fibres is placed in front of a row of photomultiplier tubes in which the enlarged shadows of the droplets produce electrical pulses of magnitude proportional to the number of fibres obscured and therefore to the droplet diameter. Since the depth of the field of view increases with the square of the particle size, the sampling volume increases with increasing particle size and so compensates for the decrease in concentration. The pulses from the photodetectors are sorted electronically into ten or more size groups corresponding to the same number of drop sizes, the number in each group being counted

---

* The slides may be prepared by wafting *clean* glass slides over burning magnesium ribbon. The preparation of a uniform layer of the correct thickness and determination of the ratio of crater size to droplet size is suggested as an exercise for the reader. It is not as easy as it sounds.

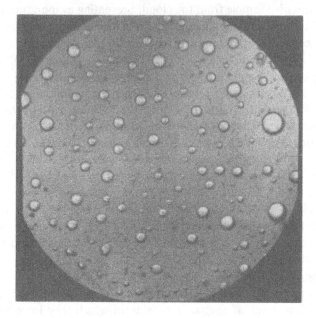

Plate VIII. A typical sample of cloud droplets caught on an oiled slide and photographed under the microscope in an aircraft. The largest drop has a diameter of about 30 μ. (By courtesy of the Director-General, Meteorological Office, Bracknell.)

automatically. Droplets of diameter $> 5\ \mu$ can be measured at high sampling rates without disturbing the sample.

In another airborne device, the droplets are sucked into a small orifice at near sonic velocity and then impact and splash on an electrostatic probe held at a potential of about 500 V to produce voltage pulses that are proportional in amplitude to the size of the droplet. The instrument collects droplets of $r > 4\ \mu$ with 100 % efficiency and appears to work reliably on an aircraft with little maintenance.

Optical methods capable, in principle, of giving the size distribution of droplets without actually disturbing the cloud and of eliminating tedious drop-by-drop measurements are, of course, attractive. Techniques have been developed which involve measurement of the distribution of light intensity within a diffraction pattern set up by the cloud droplets, and also the transmission of infra-red light through the cloud at different wavelengths. Unfortunately the interpretation of the results is difficult, particularly as the methods do not give entirely unique and unambiguous answers.

The development of a reliable, accurate and labour-saving technique for cloud-droplet measurement remains an urgent requirement for cloud physics research.

When the droplet samples have been collected and measured, the size distributions may be represented in a number of ways to accentuate different features. The concentrations of droplets of different sizes are conveniently displayed as in fig. 9 which shows $n_r dr$ against r,

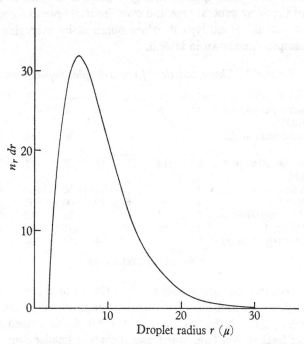

Fig. 9. An idealized distribution of cloud-droplet sizes in a small cumulus formed in continental air. The total number of drops in 1 cm³ of the cloud is given by the area under the curve and, in this case, is 300/cm³.

where $n_r dr$ is the number of droplets per unit volume of cloud of radii between r and $r + dr$. The characteristics of the distribution can be expressed numerically in terms of a number of parameters, for example:

r_m The mean radius, i.e. the sum of all the droplet radii divided by the total number of droplets.

r_{50} The median radius; half the droplets have radii smaller (or larger) than this value.

r_d The mode radius, i.e. the radius corresponding to the maximum number of droplets.

$r_{min.}, r_{max.}$ The minimum and maximum radii of the droplets in the sample.

n The total number of droplets in 1 cm³ of cloudy air.

w The total mass of the droplets in unit volume of air; this is known as the liquid-water content and usually expressed in g/m³.

The values of these quantities vary a good deal from cloud to cloud of the same general type and even from sample to sample in the same cloud. Some typical values obtained by averaging over many samples are shown in table 3.

TABLE 3. *Characteristics of cloud-droplet populations*

Cloud type	$r_{min.}(\mu)$	r_{50}	r_d	r_m	$r_{max.}(\mu)$	n (cm⁻³)	w (g/m³)
Small continental cumulus (Australia)*	2·5	6	—	—	10	420	0·40
Small continental cumulus (England)*	—	—	4	6	30	210	0·45
Small trade-wind cumulus (Hawaii)*	2·5	10	11	15	25	75	0·50
Cumulus congestus	3	—	6	10	50	100	1·0
Cumulonimbus	2	—	6	20	100	100	2·0
Orographic cloud (Hawaii)	5	13	—	—	35	45	0·30
Stratus (Hawaii)	2·5	13	—	—	45	24	0·35
Stratocumulus (Germany)	1	—	3·5	4	12	350	—

* Not more than 7000 ft deep.

At present the data are too scanty to make detailed comparisons between the characteristics of the droplet populations of different clouds very profitable, but table 3 indicates that clouds formed in the middle of large oceans (e.g. near Hawaii) contain smaller concentrations of (larger) droplets than those formed in continental air and having much the same concentrations of liquid water. This is obviously a consequence of the higher nucleus content of continental air. There is a tendency in cumuliform clouds, as they progress from small cumulus through cumulus congestus to cumulonimbus, for droplet numbers to fall but the average droplet size to increase.

Although many of the published measurements of droplet-size distributions from non-precipitating clouds can be represented by curves of the general shape shown in fig. 9, recent data indicate that droplet spectra obtained from even small cumulus clouds are often markedly bimodal with peaks at droplet radii of about 5 μ and 15–20 μ

and that the concentration of small droplets ($r < 5\,\mu$) does not fall off so sharply as indicated in fig. 9. The importance and explanation of these features will appear later.

The liquid-water content of clouds

The concentration of liquid water in a cloud, however divided among the cloud droplets and raindrops, is of considerable interest. Its magnitude and spatial distribution are important factors in the study of cloud development since they indicate the degree of mixing which has taken place between the rising cloudy air and its drier environment. Moreover, changes in water content are accompanied by large energy changes; the condensation of 1 g of water, liberating 600 cal of latent heat, would warm 1 kg of air (about 1 m³ at the lowest levels), by 2·5 °C. More detailed knowledge of its magnitude and distribution in the cloud would enable us to make more precise calculations on the growth rates of raindrops and hailstones (see pages 77–95), and a better evaluation of the icing risks to aircraft.

But again we encounter great experimental difficulties and have, as yet, no entirely satisfactory method for obtaining a continuous, easily interpreted record of the liquid-water content of both super-cooled and non-supercooled clouds over the range 0·05 g/m³ to about 5 g/m³, with an accuracy of 10 %. The instrument should also have a rapid response of not more than a second or two so that it can follow fluctuations on the scale of 100 m in the cloud.

It is, of course, possible to evaluate the liquid-water content of a cloud by adding up the volumes of all the droplets caught from a known volume of air. This method is not only laborious but its accuracy depends very much on the correct sampling and counting of the rare largest droplets which, although few in number, may contribute appreciably to the water content.

In supercooled clouds, in which the droplets freeze after hitting the collector, the water content may be determined from the rate at which the ice builds up, a knowledge of the collecting area, and the airspeed. Sometimes the collector is heated, the minimum electrical power necessary to keep it dry and free from ice being a measure of the rate of arrival of liquid water, whether supercooled or not. The power setting of the heater may therefore be calibrated, for a given airspeed, in terms of liquid-water content. Alternatively, the collector may take the form of a heated wire. The impinging and grazing droplets evaporate, cool the wire, and thereby change its electrical resistance. Such a change may disturb the balance of a Wheatstone Bridge, the out-of-balance current affording a measure of the liquid-water content. An interesting technique for use in non-supercooled clouds has been developed in Australia. The cloud droplets impinge on, and are absorbed by, a moving paper tape. Changes in the water content of the paper are detected by measuring its electrical resistance, the values of which can be calibrated in terms of the water content of the cloud. This instrument has a lag of less than 1 s and an overall accuracy of about 25 %. In order to convert the readings of all these instruments to values of cloud-water content, they have to be calibrated in a wind tunnel using an artificially produced water cloud.

Although for such purposes as estimating the danger of ice accretion on aircraft, a measure of the average liquid-water content over a flight-path of some miles may be adequate, for a detailed study of cloud structure we should like, ideally, to obtain continuous records of both water content and cloud-droplet size at several different levels simultaneously in order to obtain an instantaneous three-dimensional picture to be correlated with similar information regarding the temperature and air motions. Because of the great expense, to say nothing of the difficulties and danger of flying several aircraft simultaneously in the same cloud, it is not surprising that such comprehensive information has not yet been obtained. We have only the data provided by single aircraft making *successive* passages through a cloud at different levels with the limitation that the structure and development of the cloud, especially in the case of short-lived cumulus, may well have greatly changed between the time of the first and last traverses. Nevertheless, the measurements show a number of important characteristics, the most obvious of which is the extreme variability encountered from one sample to the next. Both the liquid-water content and the mean droplet size of a cumulus are small near the cloud base, increase with increasing height, and then decrease again to reach low values again near the cloud summit. The *maximum* concentration of liquid-water to be expected at any level, in the absence of precipitation, may be calculated on the assumption that the air rises from the cloud base and cools adiabatically without being diluted by mixing with the surrounding air. The measurements show that, at all levels, the water content w is nearly always less than this maximum or adiabatic value w_a. In small cumulus, less than about 7000 ft (2 km) deep, the value of w/w_a is, *on average*, about 0·3 near cloud base and decreases to about 0·1 at 5000–6000 ft. However, much higher values, occasionally approaching the adiabatic value, are sometimes found in localized regions of the cloud perhaps only 100 m or so in extent. Values exceeding the adiabatic are sometimes recorded in *precipitating* clouds due to local concentrations of rain water. The measurements of liquid-water content afford evidence for the fact that the rising thermals of cloudy air are diluted by mixing with their drier environment as described in chapter 1.

The average water content of small cumulus is usually only a few tenths of a gramme per cubic metre and rarely exceeds 1 g/m³; about half the traverses reveal values of $w/w_a < 0·3$. In large cumulus congestus, values greater than 1 g/m³ are often recorded, and even higher concentrations become more frequent as the cloud evolves towards a cumulonimbus.

Layer clouds, such as altostratus, stratocumulus and stratus, usually contain only small concentrations of liquid water, ranging mostly from 0·05 to 0·50 g/m³, although higher values are occasionally experienced.

The theory of droplet growth

There are two possible processes by which nucleus droplets may grow to radii of several microns and so form a cloud or fog: (i) the diffusion of water vapour to, and its subsequent condensation upon, the surfaces of the droplets; and (ii) growth by the collision and coalescence of droplets moving relative to each other due to their different rates of fall under gravity, to small-scale turbulent motions in the air, Brownian motion, or electrical forces.

(i) Droplet growth by condensation

Let us first consider the growth of a stationary, isolated, single droplet about a hygroscopic nucleus in an infinite atmosphere maintained at constant temperature and supersaturation.

The rate of increase of mass of a droplet of radius r is given by

$$\frac{dm}{dt} = 4\pi r D(\rho - \rho_r), \tag{3.1}$$

where D is the diffusion coefficient of water vapour in air, ρ is the vapour density at distances remote from the droplet and ρ_r the corresponding value at the surface of the droplet.

Since

$$\frac{dm}{dt} = 4\pi r^2 \rho_L \frac{dr}{dt}, \quad \text{and} \quad \rho = \frac{pM}{RT},$$

where ρ_L is the density of water, p the vapour pressure, M the molecular weight of water, R the universal gas constant, and T the absolute temperature of the air, (3·1) may be re-written as

$$r\frac{dr}{dt} = \frac{DM}{\rho_L RT}(p - p_r') = \frac{DM}{\rho_L RT}(Sp_\infty - p_r'), \tag{3·2}$$

where p_r' is the vapour pressure at the droplet surface, p_∞ the saturation vapour pressure at the air temperature T, and $S = p/p_\infty$ is the saturation ratio of the environment.

The growth rate of the droplet is controlled not only by the rate at which water vapour can diffuse to its surface, but by the rate of condensation, which is limited by the rate at which the liberated heat of condensation can be dissipated. Nearly all this heat is lost from the droplet surface by conduction through the air. The equation describing the heat transfer from the droplet is very similar in form to (3·2) and may be written

$$r\frac{dr}{dt} = \frac{K}{L\rho_L}(T_r - T), \tag{3·3}$$

K being the thermal conductivity of the air, L the latent heat of condensation and T_r the surface temperature of the droplet which is higher than that of its surroundings.

We also have (2·3) for the equilibrium vapour pressure p_r' over the surface of a solution droplet

$$\frac{p_r'}{p_\infty} = 1 + \frac{2\gamma M}{\rho_L RTr} - \frac{8\cdot 6m}{M_1 r^3}, \tag{3·4}$$

and the Clausius–Clapeyron equation for the variation of the saturation vapour pressure with temperature

$$\frac{1}{p_s}\frac{dp_s}{dT} = \frac{LM}{RT^2}. \tag{3·5}$$

By making only slight approximations, the last four equations may be combined to give*

$$r\frac{dr}{dt} = \frac{\left[(S-1) - \frac{2\gamma M}{\rho_L RTr} + \frac{8\cdot 6m}{M_1 r^3}\right]}{\left[\frac{L\rho_L}{KT}\left(\frac{LM}{RT} - 1\right) + \frac{\rho_L RT}{DMp_\infty}\right]} \tag{3·6}$$

* See *Physics of Clouds*, pp. 122–3.

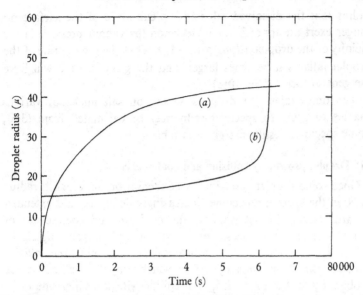

Fig. 10. Curve (a): the radius of a droplet as a function of time when growing on a salt nucleus of mass 10^{-12} g under a constant supersaturation of 0·05 % and temperature 273 °K. Curve (b): the growth of a droplet by coalescence in a cloud of liquid-water content 0·8 g/m³ comprised of droplets of radius 15 μ.

TABLE 4. *Rate of growth of droplets by condensation on salt nuclei*

Temperature, T = 273 °K　　　　　　　Pressure = 900 mb

Nucleus mass m ...	10^{-14} g	10^{-13} g	10^{-12} g
Supersaturation = $100(S-1)$ %	0·05	0·05	0·05

Radius (μ)	Time (s) to grow from initial radius of 0·75 μ		
1	2·4	0·15	0·013
2	130	7·0	0·61
5	1000	320	62
10	2700	1800	870
20	8500	7400	5900
30	17500	16000	14500
40	44500	43500	41500

which shows how the growth rate of the drop is determined by the size and nature of the nucleus, the supersaturation of the air, the rate of diffusion of water vapour to, and the conduction of heat from, the droplet. When the droplet has grown to such a radius that the dissolved salt and the curvature of the surface no longer exert an appreciable influence on the vapour pressure in the vicinity of the droplet, $(dr/dt) \propto (S-1)/r$. Thus the growth rate of the droplet falls as it becomes larger, and the growth curve will have the general form of fig. 10(a).

The times taken for droplets arising on salt nuclei of various masses to grow to specified radii may be calculated from (3·6); some specimen results are given in table 4.

(ii) Droplet growth by collision and coalescence

Once some droplets growing by condensation approach a radius of 20 μ, the spectrum becomes increasingly deformed and extended towards larger droplet sizes by the collision and coalescence of droplets of dissimilar sizes. The growth of these larger droplets by collision with smaller ones is critically dependent on (i) *the collision cross-section* or *collision efficiency*, which may be defined as the probability that a larger drop will collide with a smaller one in its direct path, and (ii) the *coalescence efficiency*, defined as the fraction of colliding drops that coalesce. The product of these two parameters, which is a sensitive function of drop size, determines the growth rate of the drops and is termed the *collection efficiency*. It is this latter quantity which is measured in laboratory experiments which, so far, have not been able to determine collision and coalescence efficiencies separately. Collision efficiencies can, in principle, be calculated theoretically, the problem being to compute the relative motions or trajectories of two approaching drops under the combined action of gravitational, hydrodynamical and electrical forces, to determine grazing trajectories, and thereby the collision cross-section of a drop for all smaller droplets. The situation is illustrated in fig. 11, where AB represents the trajectory of the centre of a droplet of radius r just making grazing contact with the drop of radius R. The effective collision cross-section of the drop for the droplet is πY_c^2, where Y_c is the initial horizontal separation of the centres on the grazing trajectory. The collision cross-section is then defined as

$$E' = \pi Y_c^2/\pi R^2 = Y_c^2/R^2, \qquad (3·7)$$

while the collision efficiency is given by—

$$E = \pi Y_c^2/\pi(R+r)^2 = Y_c^2/(R+r)^2 \qquad (3·8)$$

and has a maximum value of unity. Of course $E \to E'$ as $r/R \to 0$.

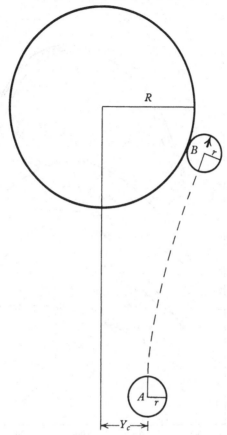

Fig. 11. Diagrammatic representation of the trajectory of a small drop relative to a larger one.

Calculations of the collision efficiencies for drop–droplet pairs falling under gravity, especially when the drop and droplet are of comparable size and when their flow patterns mutually interfere, are difficult. No complete mathematical solution for this two-body problem has yet been achieved but approximate solutions for the limited case when both drop and droplet are small enough to obey Stokes's Law (i.e. for $R \leqslant 30\,\mu$) have been obtained and the results are summarized in fig. 12. When the drop is much larger than the droplet the calculations are much easier, and computed collision efficiencies for larger drops are listed in the appendix. The most important conclusion of Hocking's original calculations was that drops of $R < 19\,\mu$ have zero collision efficiency for all smaller

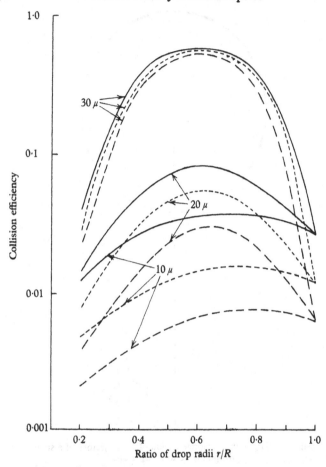

Fig. 12. Computed values of collision efficiences for small droplets of radius *r* and drops of radius *R*. (From Jonas, *Quart. J. Roy. Met. Soc.* **98** (1972), 681.)

droplets; although the later calculations show that this is not strictly accurate, the collision efficiencies for drops of *R* = 10–20 *μ* are only of the order of 1 % and do not change the essential conclusion that drops must attain radii of 20 *μ* by condensation before they can grow at an appreciable rate by coalescence. Nearly all treatments of drop growth by coalescence published before 1965 treated it as a continuous process in which the small droplets, of constant size and number, were visualized as filling space with a uniform density of liquid water which the large drops sweep up continuously. On this 'continuous' model all the large drops grow at the same rate given by

$$\frac{dR}{dt} = \frac{\pi}{3} \int_0^R n(r) r^3 E'(R,r) [V(R) - v(r)] \, dr, \qquad (3 \cdot 9)$$

where $n(r) \, dr$ is the number of droplets per unit volume with radii between r and $r + dr$, and $E'(R,r)$ is the collection cross-section for two droplets of radii R and r, with terminal velocities $V(R)$ and $v(r)$ respectively. In the simple case where all the small droplets have the same size and together constitute a liquid-water content of w grammes per unit volume of air, (3·9) reduces to

$$\frac{dR}{dt} = \frac{E'(R,r)}{4 \rho_L} w [V(R) - v(r)], \qquad (3 \cdot 10)$$

where ρ_L is the density of liquid water. The growth of a drop from a radius of 15 μ to 60 μ in a cloud containing 0·8 g/m³ of liquid water composed of droplets of radius 15 μ is shown in curve (b) of fig. 10. Comparison of curves (a) and (b) reveals that the growth rates of the two droplets, one growing by condensation on a salt nucleus of mass 10^{-12} g under a supersaturation of 0·05 % and the other by coalescence, become equal at droplet radii of $\sim$ 25 μ but, thereafter, the growth rate by coalescence accelerates while that due to condensation levels off. Of course, in a cloud, droplets grow both by condensation and coalescence acting simultaneously and there is no sudden discontinuity at which coalescence takes over from condensation. This has important consequences for the production of droplets of radius > 25 μ beyond which they may grow rapidly by coalescence into raindrops and this will be discussed in more detail later.

There has been much discussion in the literature on the possibility of droplet growth being significantly accelerated in the presence of electric charges and fields. However calculations show for example that the collision efficiency for a pair of drops of radii 20 μ and 10 μ is not significantly increased unless the electric field exceeds 100 V/cm, when it attains a value of 0·08 compared with 0·02 in zero field. A comparable effect would be achieved by electrostatic attraction between a charged and an uncharged drop if the former carried a charge of order 10^{-6} e.s.u. These conditions are likely to prevail only in incipient thunderstorms when precipitation is already well under way. Accordingly it seems that electrical forces are unlikely to accelerate the early growth of droplets by coalescence.

The growth of a population of cloud droplets

Having discussed the growth of a single droplet, we now consider the growth of a droplet population leading to the formation of a cloud. The evolution of droplets growing on a population of condensation nuclei while being lifted at an arbitrarily assigned vertical speed in air which cools adiabatically *without* mixing with its surroundings has been computed, in slightly different fashion, by several different workers. Having specified the initial temperature, pressure, humidity of the air and its nucleus content, the problem is defined by the following equations:

The equation of droplet growth:

$$r\frac{dr}{dt} = \frac{\left[(S-1)-\frac{2\gamma M}{\rho_L RTr}+\frac{8\cdot6m}{M_1 r^3}\right]}{\left[\frac{L\rho_L}{KT}\left(\frac{LM}{RT}-1\right)+\frac{\rho_L RT}{DMp_\infty}\right]}. \tag{3.6}$$

The rate of change of supersaturation

$$\frac{dS}{dt} = AU-B\frac{dw}{dt}, \tag{3.11}$$

where

$$\frac{dw}{dt} = \tfrac{4}{3}\pi\frac{\rho_L}{\rho_a}\frac{d}{dt}\overset{r}{\sum}nr^3, \tag{3.12}$$

U is the vertical velocity of the air, w the liquid-water content of the cloud in grammes per gramme of air, composed of n droplets per cm^3 of radius r, and ρ_a is the air density. A and B are parameters involving functions of temperature and pressure only. (At $T = 283\ °K$, pressure 900 mb, $A = 5\cdot36\times10^{-6}$ cm^{-1}, $B = 385$.)

Equation (3.11) expresses the fact that the supersaturation is determined by the rate at which water vapour is released for condensation by lifting and cooling of the air minus the rate at which it is condensed on to the droplets.

The rate of change of temperature in the rising cloud mass is given by

$$\frac{dT}{dt} = -\frac{g}{c_p B}U\left(\frac{P}{\epsilon p_\infty}+\frac{LM_a}{RT}\right), \tag{3.13}$$

where g is the acceleration due to gravity, P, c_p the total pressure and specific heat of the air, $\epsilon\ (= 0\cdot622)$ the specific gravity of water vapour relative to that of dry air, and M_a the molecular weight

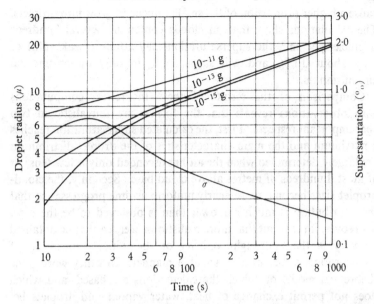

Fig. 13. The growth of cloud droplets in air which rises steadily at 1 m/s above cloud base without mixing with its surroundings. The air contains ten groups of salt nuclei ranging in mass from 5×10^{-16} g to 10^{-11} g, the concentrations being inversely proportional to their masses and ranging from 3.8×10^4 per g of air for the smallest to 1.9 per g of air for the largest. Growth curves are shown for only three nucleus classes. The curve marked σ shows how the supersaturation of the air varies with time; it attains a maximum value of 0.66% after 20 s.

of dry air ($= 28.9$). These four equations may be solved to give the drop-size, supersaturation, liquid-water content and temperature in the cloud as functions of time and the corresponding heights above cloud base. The results of a typical calculation are shown in fig. 13, which has a number of interesting features.

It demonstrates that, as cooling of the air proceeds, the supersaturation rises until it reaches a maximum value beyond which the water vapour is being absorbed by the growing drops faster than it is being released by cooling; in consequence, the supersaturation falls again. Those nuclei whose critical supersaturation for growth (see fig. 4) are exceeded develop into cloud droplets.

The droplets grow slowly at first until the supersaturation reaches the critical value, then they grow rapidly while the supersaturation continues to rise but thereafter more slowly as the supersaturation falls again. Higher rates of cooling, accompanying stronger upcurrents in the cloud, produce large peak values of supersaturation and cause a higher proportion of the smaller nuclei to grow into droplets. The calculations show that, in clouds containing several hundreds of nuclei per cm³, the supersaturation may attain a peak value of only about 0·1 %, and will surpass 1 % in only rather unusual circumstances.

Comparison of the results of these theoretical computations with observations reveals that the former are unrealistic in two very important respects. First, the calculated supersaturation reaches a maximum and the main characteristics of the droplet distribution are largely determined when the air has ascended only a few tens, or at most, hundreds of metres above cloud base. Secondly, the cloud-droplet sizes become more nearly uniform as time progresses so that the distribution is much narrower than is observed to be the case. Moreover, it turns out that more realistic answers cannot be obtained by varying the updraught velocity or the initial distribution of condensation nuclei in a reasonable manner. In other words, the theoretical model on which the calculations are based, and which does not permit exchange of heat, water vapour and droplets between the cloudy air and the drier surroundings, is inadequate and must be abandoned.

As we have seen, there is adequate evidence to suggest that a rising cloud mass continually mixes with its environment, so that heat, water vapour and droplets will be transferred across its boundaries by the turbulent air motions. It is therefore unrealistic to think of all the cloud droplets being retained in the cloud during its whole history. It seems that not only are the temperature, water content and vertical motions in the cloud largely controlled by the rate of mixing of the cloud with its environment, but that the size and size distribution of cloud droplets is also largely governed by the time which they spend in the cloud before being transferred to its boundaries, where they evaporate. By incorporating this idea into more complicated forms of (3·6) to (3·13) which allow for turbulent mixing of the cloud with the dry surrounding air, and also the fact that the droplets may grow by coalescence as well as by condensation, it is possible to derive droplet spectra for both stratus and cumulus clouds which resemble those which are actually measured. In particular, if a cumulus cloud is modelled as a succession of thermals ascending through, and mixing with, the residues of their predecessors, rather than as a single isolated parcel of air, it is possible to produce computed distributions of vertical velocity, liquid-water content, and droplet-size with the bimodal features mentioned on page 44, that agree quite well with observations. In model maritime clouds containing small concentrations of droplets, the spectra broaden quite rapidly and produce droplets of $r = 25\,\mu$ by condensation on nuclei of $m = 10^{-11}$ g in concentrations of order $100/m^3$ within half an hour, beyond which size they may continue to grow rapidly by coalescence to precipitation size.*

The simple treatment of droplet growth by coalescence based on (3·10) calculates only the averaged and smoothed rate of growth of all droplets of a certain radius but, in fact, growth by coalescence is a discrete step-wise process in which some of the larger cloud droplets may undergo more than the average number of chance collisions (while others undergo less), and this small statistically fortunate proportion will grow faster than the rest and produce a spread in the size distribution with a 'long tail' towards larger sizes. Telford was the first to point out the essential difference between the 'continuous' and 'stochastic' models, particularly in relation to the rate of production of a few large drops. Using a simple cloud model containing initially only two discrete sizes of drop and assuming a collision efficiency of unity for all sizes, he showed that the stochastic model produced a few large drops of radius $45\,\mu$ about six times faster than the average rate given by the continuous model. More detailed and realistic computations, using the collection parameters of fig. 12 and the appendix show that once droplets of $r > 25\,\mu$ appear in concentrations of order $100/m^3$ in a cloud, drizzle or rain may develop quite rapidly. Thus light rain may appear inside a small maritime cumulus cloud after a further 20 min. However, in a cloud of comparable size and vigour forming over land, the much larger concentrations of nuclei lead to higher concentrations of smaller droplets and it is difficult to produce a significant number of droplets of the critical $25\,\mu$ radius by condensation especially in the absence of hygroscopic nuclei of $m > 10^{-11}$ g; moreover, any that were present would grow rather slowly by coalescence with the smaller droplets because of the low collision efficiencies.

But, it is important to realize that in a cloud droplets grow by both condensation and coalescence acting simultaneously and there is no sharp discontinuity at say $25\,\mu$ radius where condensation ends and coalescence begins. Recent calculations† show that continued

* Mason and Jonas, *Quart. J. Roy. Met. Soc.*, **100** (1974), 23.
† Jonas and Mason, *Quart. J. Roy. Met. Soc.*, **100** (1974), 268.

condensation can significantly affect the growth rate of drops of $r > 25\,\mu$ by enhancing the growth of the smaller droplets which are then captured more efficiently by the larger ones. The overall result is to produce more rapid broadening of the droplet spectrum and more rapid growth of the largest drops than would occur by condensation and coalescence acting separately with an apparent gap at 20–$30\,\mu$ radius. The computations indicate that a small maritime cumulus only about $1\cdot5$ km deep and containing < 1 g/m^3 of liquid water can produce light rain if it lasts for more than half an hour and within 10 minutes of producing droplets of radius $25\,\mu$. The development of rain in continental clouds of a similar size but containing higher concentrations of smaller droplets may well require twice as long unless the collision efficiencies are enhanced by small-scale turbulence in the cloud as some laboratory experiments suggest.

4

THE GERMINATION AND GROWTH
OF SNOW CRYSTALS

Introduction

We have already seen that supercooled clouds are a common occurrence in the atmosphere, and that the co-existence of ice crystals and supercooled water droplets is germane to the release of precipitation. The initiation and development of the ice phase therefore form an important part of our story.

Although large quantities of water such as lakes and ponds do not supercool appreciably, cloud droplets commonly exist in the supercooled state down to temperatures as low as $-20\,°C$ and, on occasion, down to $-35\,°C$, while droplets of very pure water, only a few microns in diameter, may be supercooled to $-40\,°C$ in the laboratory. At temperatures below $-40\,°C$ such small droplets freeze automatically (or spontaneously) but, at higher temperatures, they can freeze only if they are infected with special tiny foreign particles called *ice nuclei*. The stability of natural supercooled clouds at temperatures above about $-15\,°C$ speaks for the rarity of *efficient* ice-forming nuclei in the atmosphere. Considerable interest therefore centres on the nature, origin and mode of action of these particles but, before raising these matters, it seems appropriate to review the experimental data on the supercooling and freezing of water when foreign nuclei are involved.

The supercooling of water containing foreign particles

Interest in the supercooling and freezing of water, matters which have been extensively studied since the early work of Fahrenheit in 1724, has greatly increased during the last fifteen years, mainly because of their essential importance in the physics of clouds and precipitation.

In cloud physics we are concerned with the temperatures at which airborne drops, varying in diameter from a few microns, to about 5 mm for the largest raindrops, will freeze, and how the attainable degree of supercooling may depend upon the drop size, the rate of cooling and the purity of the water.

Although there was some indication in the extensive writings of earlier scientists that the attainable degree of supercooling tends to increase when the volume of the water sample is reduced, there was so much variability in the results, with serious discrepancies between those of different workers, that no clear-cut relationships could be deduced. It appears that the earlier work may have failed to provide the required information for three main reasons. First, the water samples used by the different investigators varied greatly in their origin and purity; secondly they were usually contained in glass tubes or supported as drops on variously treated metal surfaces so that freezing may often have been initiated by the solid boundaries; thirdly, in any one investigation the volume of the sample was not usually varied sufficiently to establish clearly how this might be related to the degree of supercooling, particularly as there was usually a considerable spread in the freezing temperatures recorded for specimens of the same volume.

For these reasons, the whole subject has been examined afresh in recent years. A considerable improvement in the technique of investigating the supercooling of water was made in the author's laboratory by E. K. Bigg, who eliminated the influence of solid supporting surfaces by suspending the water drops at the boundary between two immiscible liquids having different densities, where they were also protected from infection by airborne particles. He also investigated a wider range of drop sizes—varying in diameter from about 20 μ to 2 cm—and thus volumes which differed by a factor of 10^9. The use of five pairs of supporting liquids, the members of a pair being practically immiscible with water and with each other, established that the observed freezing temperatures of the drops were a property of the water and not of the surrounding media.

Bigg determined the freezing temperatures of large numbers of drops of various sizes, cooled at a constant rate. Figure 14 shows the distribution of freezing temperatures of more than 1000 drops each of 1 mm diameter. The drops were made from distilled water from which gross impurities had been removed but which still contained very small foreign particles. The most frequent freezing temperature was -24 °C with half of the drops freezing below this temperature. Thus, if a large volume of water is subdivided into many smaller samples of each volume, the freezing temperatures of the latter show a simple probability distribution as illustrated in fig. 14. Because of the statistical character of the nucleation events, it is necessary to determine the freezing-points of large numbers of samples in order to obtain characteristic and significant relationships. Such an important relationship is revealed when one plots the *median* freezing temperature of a large group of drops, i.e. the temperature below which half of the drops freeze, against the

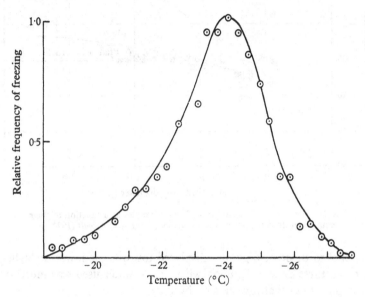

Fig. 14. The distribution of freezing temperatures of water drops of 1 mm diameter containing foreign particles. (After E. K. Bigg.)

logarithm of the drop diameter (or volume). This produces the straight-line relationship shown in fig. 15 which may be represented by the equation.

$$\log V = A - B(273 - T), \tag{4.1}$$

where V is the drop volume, T the freezing temperature in °K, and A and B are constants for the particular sample of water under test.

Bigg's work has recently been checked and extended. The results obtained with water varying in purity from that of rain water to that produced by multiple distillation showed the same general trend; plots of the median freezing temperatures of groups of drops against the logarithm of their diameters produced straight lines

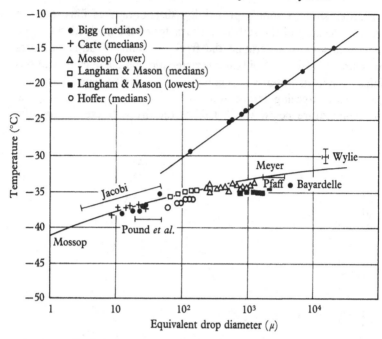

Fig. 15. The freezing temperatures of water samples as a function of their equivalent drop diameter. (From Mason, *Physics of Clouds* (1971).)

parallel to those of Bigg, but displaced towards the lower or higher temperatures depending upon whether the water used was more, or less, pure than that investigated by Bigg.

Although in these experiments the grosser particles were removed from the water, it still contained large numbers of small particles which can be removed only by taking extreme measures. Equation (4·1) therefore represents nucleation by foreign particles—heterogeneous nucleation. This relationship between droplet volume and degree of supercooling may be explained on the assumption that the water was contaminated by particles which were a typical sample of atmospheric aerosol, the freezing-nucleus content of which is observed to increase exponentially with decreasing temperature as described below.

When foreign particles are entirely excluded from the water, as in the case of the tiny droplets produced by condensation in clean air of the expansion chambers, nucleation can occur only by small groups of water molecules becoming arranged by chance microscopic thermal fluctuations into an ice-like configuration. The probability of an aggregate reaching a given size increases as the temperature is lowered until eventually it surpasses a critical size beyond which it can continue to grow with a decrease of free energy, and form a nucleus for the ice phase. Theoretical expressions for the rate of formation of such nuclei in unit volume of water, and therefore the probability of a drop of given volume freezing within a given time, have been developed but they contain parameters which cannot be accurately calculated or measured so that absolute nucleation rates cannot be reliably estimated. However, if the temperature below which a given fraction of droplets of a given size freeze within a given time is measured, the theory allows one to estimate the temperatures at which droplets of other sizes will freeze under similar circumstances. Thus the lower curve in fig. 15 indicates the temperatures at which drops of various diameters should freeze within 1 s if, as Mossop observed, droplets of $d \sim 1\,\mu$ freeze at $-41\,°C$ within 0·6 s. On the same diagram are plotted the results of a number of workers who have been able to supercool drops ranging in diameter from $10\,\mu$ to several mm to very near the temperature limits predicted by the theory of homogeneous nucleation. Figure 15 shows that distinctly different relationships hold for heterogeneous and homogeneous nucleation, and that these are valid for the wide range of drop sizes encountered in clouds and rain.

Ice nuclei in the atmosphere

Except at temperatures below $-40\,°C$, the ice-nucleus content of the air is of fundamental importance for the initiation of the ice phase in clouds. It is not easily measured. The most usual method involves the use of a cloud chamber in which a sample of atmospheric air is saturated with water vapour and rapidly cooled by sudden expansion. During this rapid cooling, water vapour condenses on some of the airborne particles to produce a cloud of supercooled droplets. Some of these will contain an ice nucleus, freeze, and grow into ice crystals. The technique is then to count the number of crystals glittering in an illuminated volume of the cloud, successive measurements being made at lower and lower temperatures achieved by larger and larger expansions. Because it is not easy to discern small numbers of crystals swirling about in the thick water fog, and one cannot be certain that every glittering particle is

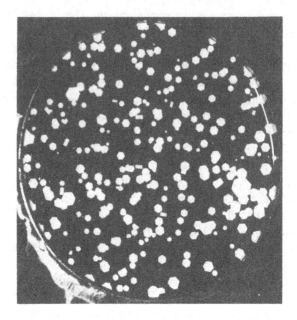

Plate IX. Supercooled sugar solution provides a convenient method of counting the tiny ice crystals created in a cloud chamber. After falling into the sugar solution, the crystals grow to a visible size of several millimetres.

an ice crystal, direct visual counts are not very accurate. This led E. K. Bigg to devise an ingenious technique in which the ice crystals fall into a tray of sugar solution placed at the bottom of the cloud chamber. The water in the solution supercools and when the tiny ice crystals fall on to it they quickly grow to visible size, as shown in pl. IX, and are easily counted.

A technique that has several advantages over cloud-chamber methods involves the drawing of large samples of air through a membrane filter, when the number of ice nuclei retained in the sub-micron pores of the filter can be determined by standardized processing in the laboratory. The filter is now impregnated with molten petroleum jelly that is allowed to block the pores but not flood the upper (collecting) surface. After allowing the jelly to set, the filter is placed on a cold metal stage in a small humidity chamber in which the air is saturated relative to the ice-coated walls maintained at, say, −18 °C. The temperature of the stage and filter is then lowered to −20 °C, and the number of ice crystals appearing on the filter counted. Growing under these conditions, in an atmosphere barely

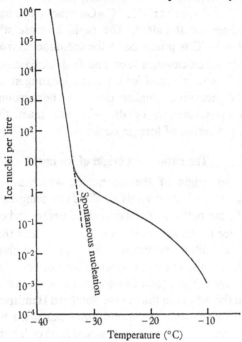

Fig. 16. The concentrations of ice crystals appearing in a cloud chamber at different temperatures. Such a curve is therefore supposed to represent the ice-nucleus content of the air, these particular measurements being representative of a day when this was rather low.

supersaturated with respect to water, many hundreds of nuclei can be activated and counted on a single filter without neighbouring crystals interfering with one another.

The overall reproducibility of the method is good and this is probably the most reliable and convenient method of determining the ice-nucleus content of the atmosphere.

Measurements made on land, at sea and in aircraft show that the ice-nucleus content of the atmosphere varies considerably from day to day and from place to place. It will therefore be a long, laborious process to obtain adequate data on the ice-nucleus content of the atmosphere and to determine how it is influenced by geographical, topographical, meteorological and other (including, perhaps, extra-terrestrial) factors.

On days when the ice-nucleus content of the air is low, the concentrations appearing in cloud chambers at various temperatures are typically those of fig. 16, with $1/m^3$ at -10 °C, $100/m^3$ at -20 °C,

1/l. at -30 °C and 1/cm³ at -35 °C. On other days, nuclei may be nearly 100 times as abundant. The rapid increase which sets in below about -33 °C is produced by the spontaneous freezing of the cloud droplets which causes a thousand-fold increase between -33 and -35 °C. The numbers of ice crystals appearing at temperatures *above* -33 °C are much higher than can be accounted for by spontaneous crystallization of droplets and must therefore be attributed to the action of foreign nuclei.

The nature and origin of ice nuclei

The nature and origin of the ice nuclei, which are responsible for the appearance of ice crystals in clouds at temperatures above about -30 °C, are matters of considerable interest and controversy.

In general, ice nuclei are more abundant over land than over the open oceans, and often, but not invariably so, more abundant near the ground than aloft. There is also some but not entirely conclusive evidence that average concentrations tend to be an order of magnitude higher in the Northern than in the Southern Hemisphere. Spatial and temporal fluctuations in concentration can often be traced to local or regional sources, for example dust storms or the effluent from industrial plants such as steelworks, and the counts may depend on the wind direction and the vertical stability of the lower atmosphere. This all seems to suggest that the nuclei are primarily of land origin and the most natural assumption to make is that they originate mainly from the earth's surface as dust particles and are carried aloft by the wind. Mason suggested that they consist largely of clay-silicate particles. Alternatively, Bowen suggested that they are of extra-terrestial origin and enter the top of the atmosphere as meteoritic dust. Schaefer found that some clay soils produced ice crystals in a cloud chamber at temperatures as high as -12 °C, and from tests on 35 rather pure and well-identified mineral dusts, Mason found 21, mainly silicate minerals of the clay and mica groups, were active nucleators to the extent that 1 particle in 10^4 produced an ice crystal at temperatures above -15 °C, and of these, 10 were active above -10 °C, the most common being kaolinite: see table 5. On the score of natural abundance and effectiveness, Mason considered the clay minerals together with illites and halloysite to be the most important sources of terrestrial ice nuclei. Supporting evidence for this view is provided by several Japanese workers, who have used the electron microscope and electron diffraction to examine the nuclei at the

TABLE 5. *Natural nuclei*

Substance	Crystal symmetry	Threshold temperature (°C)
Ice crystal	Hexagonal	0
Covellite	Hexagonal	− 5
Vaterite	Hexagonal	− 7
Beta tridymite	Hexagonal	− 7
Magnetite	Cubic	− 8
Anauxite	Monoclinic	− 9
Kaolinite	Triclinic	− 9
Illite	Monoclinic	− 9
Glacial debris		−10
Hematite	Hexagonal	−10
Brucite	Hexagonal	−11
Gibbsite	Monoclinic	−11
Halloysite	Monoclinic	−12
Volcanic ash	—	−13
Biotite	—	−14
Vermiculite	Monoclinic	−15
Phlogopite	—	−15
Nontronite	Monoclinic	−15

centres of snow crystals. Nearly 90 % of the nuclei were identified as soil particles, about 70 % being clay minerals, either kaolinite, montmorillonite, or very closely related forms. However, it is difficult to believe that such a high incidence of large clay particles is truly representative of all snow crystals, particularly if a high proportion of these originate on ice splinters (see later), and much further careful work is required to test this. On the other hand, there is little or no direct evidence that meteoritic dusts do, or can, act as efficient ice nuclei. Evidence for the meteor-dust hypothesis, which is based on correlations between the occurrence of meteor showers and periodicities in rainfall, is indirect, circumstantial, and unconvincing, and is unsupported by reasonable physical arguments on the possible magnitude and mechanism of the effect. The origin and nature of the primary ice nuclei have yet to be firmly settled.

In any case, measurements of ice nuclei in clear air may not serve as a reliable guide to the ice-crystal population of a cloud because additional nuclei may be produced inside the cloud by fragmentation of snow crystals, freezing drops, and hail pellets.

The apparent necessity for the secondary production of ice nuclei during the steady release of snow from layer-cloud systems was pointed out by Mason who showed that, in order to supply one nucleus per crystal, the air entering the cloud base would have to contain nucleus concentrations 10^2–10^3 times higher than are normally measured at $-15\,°C$, and suggested that the additional nuclei are produced by small splinters being torn off the fragile dendritic crystals and that these can then serve as nuclei for new crystals. It may be shown that an original crystal would have to shed a splinter, on average, every 30 s, about 30 splinters in all, in order to sustain the numbers concerned and to multiply the original population of nuclei one-hundredfold. This does not seem unreasonable, but the conditions under which snow crystals may splinter have not yet been investigated even in the laboratory.

Even more puzzling is that some slightly supercooled cumulus clouds, with summit temperatures only a few degrees below $0\,°C$, contain high concentrations of ice crystals—two to four orders of magnitude greater than the concentrations of ice nuclei measured in clear air at the same level. According to a series of careful and detailed aircraft observations and measurements by Mossop and his colleagues, such high concentrations of crystals occur only in cumulus that have been in existence for some time and acquired widths of several kilometres but are absent from newly-rising cloud towers. They are also absent from slightly supercooled stratocumulus clouds. Mossop reports small ice crystals in concentrations of 10–100/litre in clouds with summit temperatures as high as $-8\,°C$ but always in association with rimed ice pellets of order 1 mm diameter in concentrations of order 1/litre and with large droplets of $r > 250\,\mu$ in concentrations of order $100/m^3$. The concentrations of ice nuclei active at $-8\,°C$ are below those that can be reliably measured by conventional techniques but probably do not exceed $1/m^3$.

This large discrepancy between the concentrations of ice crystals and of ice nuclei suggests that some process of ice crystal multiplication must be active in the cloud. The most plausible suggestion is that small ice splinters are ejected during the accretion and freezing of supercooled cloud droplets on small ice pellets which would be consistent with the fact that high concentrations of ice crystals occur only in the company of relatively large rimed ice particles. Although ice splinters are sometimes observed to be ejected during the freezing of drops of radius $> 30\,\mu$ and during the formation of rime deposits, different workers have reported very different rates of splinter production, and these are generally too low to account for the high crystal concentrations reported by Mossop unless the rimed particles produced in one rising cloud tower are caught up and continue their growth in one or more successive towers.

Artificial ice nuclei

The fact that supercooled clouds are a common occurrence at temperatures above -20 °C, with the implication that efficient ice-forming nuclei are often deficient in the atmosphere, and the possibility of inducing rain by supplying artificial nuclei, have stimulated a search for substances which may be dispersed into the atmosphere in finely divided form to produce vast quantities of efficient, durable nuclei.

The most important discovery was made by B. Vonnegut in 1946, who found that silver iodide was highly effective in this respect. A dilute solution of the salt may be vaporized in a hot flame to produce about 10^{16} tiny crystals from 1 g of silver iodide. When this smoke is introduced into a supercooled cloud, some ice crystals appear when the temperature falls below -4 °C but the numbers rapidly increase with decreasing temperature until, at about -15 °C, most of the silver iodide particles serve as ice nuclei. Silver iodide, which was chosen originally because of the similarity of its atomic arrangement to that of ice, is still the most effective nucleating substance so far discovered. Its efficiency depends very much upon the method of preparation and the presence of impurities. For example, if the nuclei are produced in a reducing atmosphere, such as in a hydrogen flame, their surface layers are liable to be reduced to metallic silver and their ice-nucleating ability thereby inhibited. This effect is greatly enhanced if the crystals are irradiated with ultraviolet light of wavelength less than 4300 Å which explains why the nucleating ability of silver iodide deteriorates when dispersed into the atmosphere in strong sunshine. According to field tests, the activity of nuclei from a hydrogen burner decays to about one-hundredth of its initial value after an exposure to strong sunlight of only 15 min. Nuclei produced by a kerosene burner decay more slowly, by a factor of about 50 after an hour's exposure. This photolytic decay of silver iodide is accelerated by the presence of small traces of impurity, particularly sulphur.

Because silver iodide is rather expensive for use on a large scale, and because of its deterioration in the sunlight, an intensive search has been made for possible substitutes. A good deal of work has

TABLE 6. *Artificial inorganic ice nuclei*

Substance	Crystal symmetry	Threshold temperature (°C)
Silver iodide	Hexagonal	− 4
Lead iodide	Hexagonal	− 6
Cupric sulphide	Hexagonal	− 6
Mercuric iodide	Tetragonal	− 8
Silver sulphide	Monoclinic	− 8
Ammonium fluoride	Hexagonal	− 9
Silver oxide	Cubic	−11
Cadmium iodide	Hexagonal	−12
Vanadium pentoxide	Orthorhombic	−14
Iodine	Orthorhombic	−14

been carried out in different laboratories on the ice-nucleating ability of a wide variety of chemical compounds, but there has been little agreement in the results. Careful tests in the author's laboratory indicated that many published results are spurious because of the presence, in the air or the chemicals, of small traces of silver or free iodine, leading to the formation of silver iodide. If all such trace impurities are removed, the substances which remain as definitely active are listed in table 6. This shows the highest temperatures at which one particle in ten thousand of the substance produced an ice crystal when introduced into a supercooled cloud in the laboratory. Greater numbers of effective nuclei are obtained as the temperature is lowered below the threshold value. The first five substances listed in table 6, which are all practically insoluble in water, are active above −9 °C and are therefore all potential candidates for rainmaking experiments. However, they are all less potent and less easily dispersed than silver iodide. Ammonium fluoride, cadmium iodide, vanadium pentoxide, and iodine, being soluble in water, are inactive in a supercooled water cloud, but produce ice crystals in an environment in which the humidity is maintained between water and ice saturation at the temperature indicated.

Although there is a tendency for the more effective nucleators to be hexagonally-symmetrical crystals in which the atomic arrangement is reasonably similar to that of ice, table 6 shows that there are a number of exceptions; but, for all the substances which are active above −15 °C, it is possible to find a crystal face in which the atomic spacings differ from those in ice by only a few per cent. However, there is not, in general, a high·correlation between the

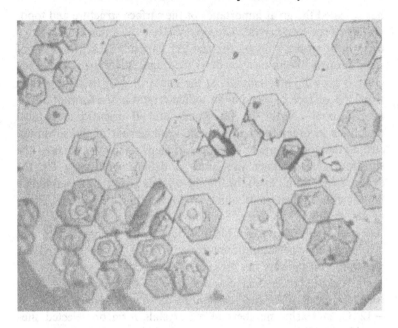

Plate x. A deposit of hexagonal ice crystals oriented parallel to one
another on the basal plane of a single crystal of silver iodide.

threshold nucleation temperature and the degree of misfit between
the ice and nucleus structure, indicating that nucleating ability is
only partly determined by crystal geometry, and that other factors,
not yet fully understood, play a role.

In an attempt to investigate the nucleation mechanism in more
detail, the author and his colleagues have studied the growth of
ice on individual faces of single crystals of various nucleating
agents, under carefully controlled conditions of temperature and
supersaturation of the vapour.

Deposits of ice crystals, all oriented parallel to one another, have
been observed on hexagonal crystals of silver iodide, lead iodide,
cupric sulphide, cadmium iodide and brucite, and also on freshly-
cleaved mica, on calcite, mercuric iodide, vanadium pentoxide and
iodine.

Plate x shows ice crystals in the form of thin hexagonal plates all
growing parallel to one another on a single crystal of silver iodide;
they appeared preferentially around local imperfections on the crystal
surface which show as darker spots in the photograph. This study

has revealed the great importance of the surface structure and topo-graphy of the host crystal. The ice crystals show a marked tendency to form at special sites on the surface, particularly at the edges of steps formed during its growth or cleavage. Plate xi(*a*) shows ice crystals growing preferentially at the edges of growth steps arising from the surface of a cadmium iodide crystal in the form of a hex-agonal spiral. Again the ice crystals are all oriented in parallel directions in order to fit the atomic arrangement of the substrate. Just as the crystals prefer the edges of steps to the flat areas of the substrate surface, they also show a definite preference for the deeper steps. In pl. xi(*b*) the few ice crystals growing on lead iodide appear only at those steps which are higher than about $0 \cdot 1 \mu$. Steps of about this height provide effective nucleation sites if the air is supersaturated relative to a plane surface of ice by about 10 %, but much higher supersaturations, exceeding 100 %, are required for nucleation on the very flat areas of the host crystal.

On silver iodide, at temperatures above $-4\,^{\circ}\mathrm{C}$, only water drop-lets are formed. As the temperature is lowered from $-4\,^{\circ}\mathrm{C}$ to $-12\,^{\circ}\mathrm{C}$, increasing numbers of ice crystals form on selected sites provided the air surpasses saturation relative to liquid water. At temperatures below $-12\,^{\circ}\mathrm{C}$, however, crystals appear when the air is under-saturated relative to water but supersaturated relative to ice by at least 12 %. Very similar results have also been obtained for lead iodide, cupric sulphide and cadmium iodide, with slightly different critical temperatures and supersaturations in each case. The inference is that on nuclei which are large enough $(r > 1\mu)$ for the curvature effect to be negligible, condensation followed by freezing occurs at special active sites at temperatures not far below $0\,^{\circ}\mathrm{C}$ but, at lower temperatures, ice forms directly from the vapour phase providing the supersaturation is high enough. In accordance with (2·1) small nuclei $(r < 0\cdot1\mu)$ require appreciable supersatura-tions for condensation to occur; if these are not achieved in natural clouds the nuclei can act only if they are captured by a supercooled cloud droplet. There is some evidence that immersion of such small particles in water droplets leads to some loss of potency—probably by the dissolution of active nucleating sites.

In addition to the inorganic crystals just described, several organic substances such as α-phenazine, metaldehyde, and a range of amino-acids and steroid compounds can act as effective ice nuclei at tempera-tures as high as $-1\,^{\circ}\mathrm{C}$. The efficacy of these substances shows no

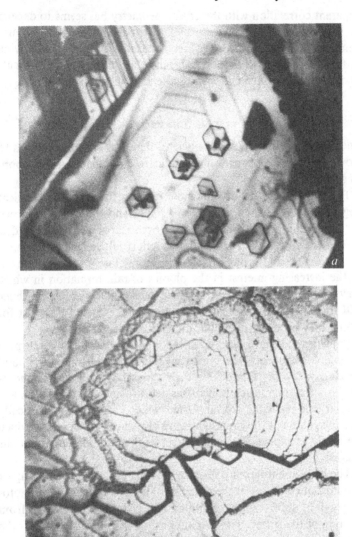

Plate XI. (*a*) Ice crystals growing on a single crystal of cadmium iodide. They form preferentially at the edges of growth steps which emanate in a spiral terrace from the cadmium iodide surface. (*b*) Oriented ice crystals originating at the edges of steps deeper than $0 \cdot 1\,\mu$ on the basal plane of lead iodide. No ice crystals appear on the shallow growth steps.

apparent correlation with the crystal geometry but seems to depend on the molecular structure especially the strength of the molecular bonding between adsorbed water molecules and the underlying surface. Because most of these substances are expensive and difficult to disperse, they have not been used extensively in cloud-seeding.

Snow crystals from natural clouds

Once germinated upon an ice nucleus, the snow crystal continues to grow by the diffusion of water vapour to, and condensation upon, its surfaces.

The remarkable beauty of snow crystals, seen in the classical elegance of the simple geometrical shapes and the delicate tracery of the more intricate forms, has long been recognized and recorded by the scientist, the artist, and the industrial designer. Recently their structure and growth have been more closely studied because of the increasing interest in the physics of rain formation in which they play an important part. Their remarkable variety of shape and form has also caught the attention of the crystal physicist for whom they pose intriguing but difficult problems.

Snow crystals have been collected, photographed and catalogued by many enthusiasts, the books by Bentley and Humphreys[*] and by Nakaya[†] being famous for their thousands of beautiful photographs. However, such collections tend to include only the most regular and striking crystal forms which, although aesthetically satisfying, are not truly representative of natural snow crystals which are rarely as symmetrical and perfect as the published photographs suggest.

The direct photography of snow crystals present difficulties apart from the obvious one of their melting. The crystals have to be photographed in subfreezing conditions when trouble may arise from frosting of the lenses; also precautions have to be taken against the crystal evaporating while being handled. These difficulties may be overcome by making permanent plastic replicas of the crystals as described on page 92.

The replicas or the crystals themselves may be photographed through a low-power microscope by either reflected or transmitted light. Transmitted light produces a clear picture of the boundary

---

[*] *Snow Crystals*, by W. A. Bentley and W. J. Humphreys (McGraw-Hill, 1931).
[†] *Snow Crystals*, by U. Nakaya (Harvard University Press, 1954).

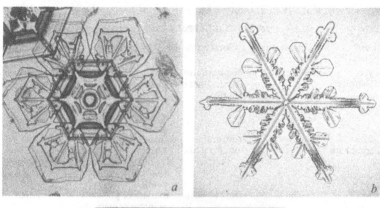

Plate XII. (*a*) A sector plate crystal photographed in transmitted light. (Photograph by courtesy of Urs Beyeler.) (*b*) A dendritic stellar snow crystal photographed in transmitted light. (Photograph by courtesy of Urs Beyeler.) (*c*) A dendritic stellar crystal photographed in reflected light. (*d*) A dendritic stellar crystal photographed by oblique illumination. (Photograph by courtesy of U. Nakaya.)

TABLE 7. *Weickmann's observations of predominant crystal
forms in different cloud types*

Level of observation	Temperature range	Cloud types	Crystal forms	Crystal sizes (approx.)
Lower troposphere	0 to − 15 °C	Nimbostratus, stratocumulus, stratus	Thin hexagonal plates	50μ to $\frac{1}{2}$ mm dia., 10 to 20μ thickness
			Star-shaped crystals showing dendritic structure	$\frac{1}{2}$–5 mm dia.
Middle troposphere	− 15 to − 30 °C	Altostratus, altocumulus	Thick hexagonal plates	$\sim 200\mu$ dia.
			Prismatic columns, single prisms and twins	$\sim 200\mu$ long
Upper troposphere	< − 30 °C	Isolated cirrus	Clusters of prismatic columns containing funnel-shaped cavities	~ 1 mm dia.
			Single hollow prisms	$\sim \frac{1}{2}$ mm long
		Cirrostratus	Single complete prisms	$\sim 100\mu$ long, length to diameter = 1–5

and the internal structure of the crystal (see pl. XII(*a*)). Reflected light produces a white image on a black background and reveals more of the surface relief (pl. XII(*c*)). The best results are obtained by the use of oblique illumination which combines the advantages of both transmitted and reflected light, and reveals both the internal and surface structures (pl. XII(*d*)).

Although extensive observations have been made of snow crystals reaching the ground, and many attempts made to correlate the relative frequencies of the various crystal forms with the temperature at the place of observation, very little in the way of consistent results has emerged. This is not, perhaps, surprising since only the conditions prevailing during the growth of the crystal are relevant. It is only in recent years that crystals have been collected from different types of cloud having widely different conditions of temperature, water-vapour content, and supersaturation relative to ice.

The aircraft observations, summarized in table 7, are not as detailed as one could wish, but the sampling of crystals from an aeroplane presents considerable difficulties, particularly at high speeds when it becomes difficult to avoid shattering the crystals during their collection. Nevertheless, they do reveal that, at a given temperature, a particular crystal type tends to be dominant.

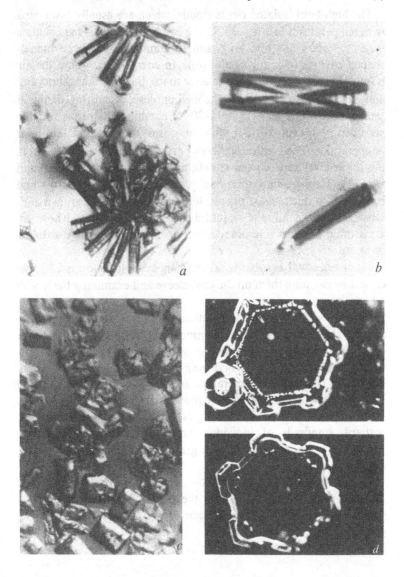

Plate XIII. (*a*) Clusters of hollow prismatic ice columns collected from cirrus clouds at −40 °C. (Photograph by H. K. Weickmann.) (*b*) Individual hollow prismatic columns from cirrus clouds. (Photograph by H. K. Weickmann.) (*c*) Solid prismatic columns from cirrostratus at −26 °C. (Photograph by H. K. Weickmann.) (*d*) Thin hexagonal ice plates. (Photograph by H. K. Weickmann.)

The high-level isolated cirrus clouds, which are usually associated with temperatures below −30 °C, are composed of six-sided columns (prisms), typically $\frac{1}{2}$ mm long, and containing pronounced funnel-shaped cavities (pls. XIII(*a*) and (*b*)). In cirrostratus, where the air is only slightly supersaturated relative to ice, the prisms are short and solid as shown in pl. XIII(*c*). The medium-level clouds, occurring at temperatures between −15 and −30 °C, contain both prisms and thin six-sided plates (pl. XIII(*d*)), with the prisms dominant at the lower temperatures. The greatest variety of crystal shapes is to be found in the lower-level supercooled clouds, at temperatures between 0 and −15 °C. Here we find hexagonal plates, perhaps $\frac{1}{2}$ mm across and only 10–20 μ thick, short prisms, long, thin needles (pl. XIV(*a*)) and, most striking of all, the beautiful star-shaped crystals, whose six arms often develop side branches to produce the fern-like patterns of pl. XII.

The reader will be able to identify these various forms of snow crystal by catching them on the coat-sleeve and examining them with a hand-lens.

Crystals which are a combination of two or more basic types (prism, plate, star) are not uncommon; the prismatic column with end-plates and the combination of plate and star crystal, shown in pls. XIV(*b*) and XII(*d*) are good examples. The metamorphic forms reflect the changes in temperature and supersaturation which the crystal experiences on its journey towards the ground.

Snowflakes, formed by the aggregation of between two and, perhaps, hundreds of individual crystals, with the stars usually prominent as shown in pl. XIV(*c*), grow at temperatures only a few degrees below 0 °C.

A classification of natural snow crystals by Magono and Lee appears in fig. 17 and shows how the shape or habit of the crystals varies with the temperature and supersaturation experienced during growth. Some typical values for the diameters, masses and falling speeds of the various crystal types are shown in table 8.

The growth of snow crystals in the laboratory

The remarkable variety of snow-crystal forms suggest that their growth and development are complicated matters which may be studied with advantage in the laboratory where the whole life history of a crystal can be observed under controlled conditions.

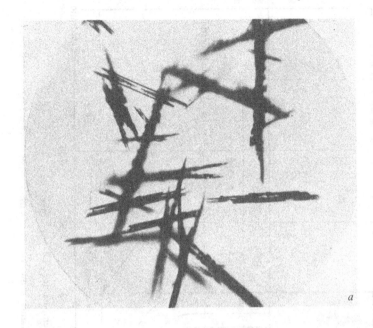

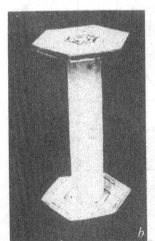
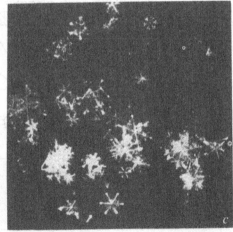

Plate xiv. (*a*) Thin ice needles. (Photograph by U. Nakaya.) (*b*) A hexagonal column capped with hexagonal end plates. (From Bentley and Humphreys, *Snow Crystals* (McGraw Hill, 1931.) (*c*) Snowflakes composed largely of stellar snow crystals.

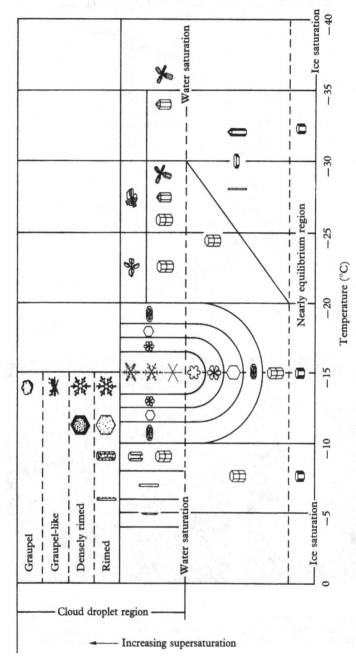

Fig. 17. Temperature and humidity conditions for the growth of natural snow crystals of various types. (After Magono and Lee.)

TABLE 8. *The diameters, masses, and fall velocities of snow crystals*

	d (mm)	m (mg)	v (cm/sec)
Needle	1·5	0·004	50
Plane dendrite	3·0	0·04	30
Spatial dendrite	4·0	0·15	60
Powder snow	2·0	0·06	50
Rimed crystals	2·5	0·18	100
Graupel	2·0	0·80	180

Small crystals have been grown in supercooled clouds produced by the introduction of steam into room-size refrigerators whose temperature may be controlled. Within a few seconds, the fog cools to the temperature of the cold chamber and it may then be seeded with either dry ice or silver iodide to produce sufficient crystals. Having grown for a minute or two in the supercooled fog, the crystals fall to the floor of the chamber where they may be caught on microscope slides and photographed under the microscope. The types of crystal which grow in this way at various temperatures are shown in table 9. Their classification according to temperature bears a marked similarity to that of natural snow crystals, showing that it is possible to simulate quite well the early stages of snow-crystal growth in the laboratory and, at the same time, to determine the temperature-ranges for the different crystal forms more precisely than can be done in the atmosphere.

TABLE 9. *Changes of crystal habit with temperature in artificially produced water clouds (after aufm Kampe et al., and Mason)*

Temperature range	Crystal habit
0 to −4 °C	Simple clear plates with no surface markings; some trigonal shapes
−4 to −9 °C	Prisms, some showing marked cavities and similarity to needles
−10 to −25 °C	Hexagonal plates showing ribs, surface markings, and tendency to sprout at corners. Sector stars. Dendritic stars, most prominent below −14 °C
−25 to −40 °C	Single prisms, twins, and hollow prisms. Aggregates of prisms and irregular crystals (aufm Kampe *et al.*)

Experiments to study the growth of individual snow crystals in different environmental conditions were first carried out by U. Nakaya and his colleagues in Japan. The crystals were grown on fine rabbit's hair stretched on a frame and suspended in an air stream whose temperature and water-vapour content could be varied. Growth of the crystal was followed by time-lapse photography. The apparatus consisted of two concentric glass cylinders, the warm water vapour from a heated reservoir being convected upwards inside the inner tube, cooled on its way up, and returned through the annular space between the cylinders. The whole apparatus was placed in a thermostatically controlled cold chamber. The degree of supersaturation in the experimental space was varied by altering the temperature T_w of the water in the reservoir. The temperature T_a of the air in the immediate vicinity of the growing crystal was determined by both T_w and the temperature of the thermostat.

Crystals were produced at various combinations of T_a and T_w but, unfortunately, the conditions in the apparatus were not steady nor well defined because the strong convection gave rise to large fluctuations in both temperature and supersaturation. Nevertheless, it emerged that T_a and T_w were the two main factors controlling the growth forms of the crystals, and Nakaya's classification of crystal shape in relation to the air temperature is in broad agreement with the data of tables 7 and 9. Nakaya was of the opinion that the temperature rather than the supersaturation of the environment was the main factor controlling the crystal shape except for the branching (dendritic) star-shaped crystals which occurred only at relatively high supersaturations and at temperatures between -14 and -17 °C.

However, other workers have interpreted Nakaya's results as showing that the crystal habit is principally determined by the flux of vapour directed towards the crystal, a quantity which is closely related to the supersaturation. Strong evidence against such an interpretation was obtained in the author's laboratory when ice crystals were grown on a metal surface under conditions such that the temperature and supersaturation of the surrounding air could be varied and measured independently. The prism and plate-like forms were produced at the same temperatures as given in table 9, and it was clearly established that whether a crystal grew as a prism or a plate was determined only by the temperature, the supersaturation, though varied over wide limits, having no systematic effect.

It must be admitted, however, that conditions for growth on a metal surface may not have fairly simulated those occurring in free air. To meet this point and to study the whole problem in much more detail, Mason and Hallett carried out a new series of experiments.

The crystals are grown on a thin nylon fibre running vertically through the centre of a diffusion cloud chamber in which the vertical gradients of temperature and supersaturation can be accurately controlled and measured. In this apparatus, sketched in fig. 18, vapour diffuses from a water source at room temperature in the top of the chamber towards the base which is cooled to about $-60\,°C$. This produces strong vertical gradients of both temperature and water vapour density. The very strong temperature inversion damps out convection and turbulence and keeps the air very still. The temperature at any level in the chamber can be accurately measured with movable thermocouples and a special technique is. used to determine the vapour density and hence the supersaturation at any level. If the chamber is filled with atmospheric air, water vapour condenses on the condensation nuclei and produces a cloud. In the top third of the chamber, the cloud is warmer than $0\,°C$. In the middle third, at temperatures between 0 and $-40\,°C$, we have a cloud of supercooled droplets and maybe a few ice crystals formed upon the natural ice nuclei in the air. In the bottom third of the chamber, at temperatures below $-40\,°C$, the cloud is composed entirely of ice crystals. If the chamber is allowed to stand for several hours, all the condensation nuclei settle out and leave clean, particle-free air in which the supersaturation may build up to $300\,\%$ at which condensation occurs on the ions produced by cosmic rays.

With this equipment, crystals have been grown on a vertical fibre over the temperature range 0 to $-50\,°C$, and under supersaturations ranging from a few per cent in the presence of a water-droplet cloud to about $300\,\%$ in clean, droplet-free air. The results of many experiments are summarized in fig. 19. Consistently the crystal habit varied along the length of the fibre in the following manner:

0 to $-3\,°C$	Thin hexagonal plates
-3 to $-5\,°C$	Needles
-5 to $-8\,°C$	Hollow prismatic columns
-8 to $-12\,°C$	Hexagonal plates
-12 to $-16\,°C$	Dendritic, fern-like crystals
-16 to $-25\,°C$	Hexagonal plates
-25 to $-50\,°C$	Hollow prisms

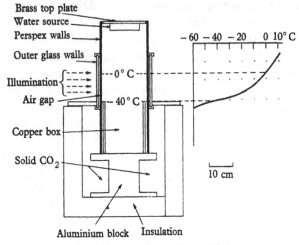

Fig. 18. The diffusion cloud chamber.

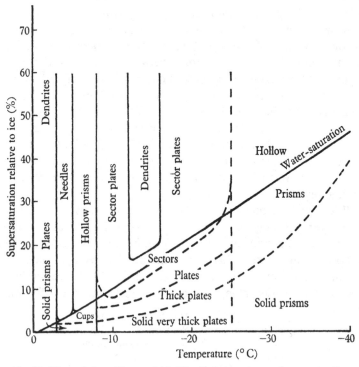

Fig. 19. The variation of ice crystal habit with temperature and supersaturation of the vapour.

This scheme is very similar to that of table 9 for crystals grown in laboratory-produced clouds, but the simultaneous growth of all the crystal forms on the same fibre brings to light the sharpness of the boundaries between one habit and another. For example, the transition between the plates and needles at -3 °C, and that between the hollow prisms and plates at -8 °C, occurred within temperature intervals of less than one degree. A photograph showing the variation of crystal form along the length of the fibre is reproduced in pl. xv.

Crystals having an almost identical variation of habit with temperature have also been grown from the vapour of heavy water (pl. xvi), but with the transition temperatures all shifted upwards by nearly 4 °C, in conformity with the melting-point of D_2O being 3·8 °C.

These experiments showed clearly that very large variations of supersaturation do not change the basic crystal habit as between prism and plate-like growth although, of course, the supersaturation does control the growth rate. Experiments in which crystals were grown at supersaturations of only a few per cent showed that the supersaturation also controls the aspect ratio of the crystal. Thus at temperatures where plate-like crystals appear, increasing supersaturation causes transitions from very thick plate → thick plate → sector plate → dendrite, and, in the prism régime, the development is from short solid prism → longer hollow prism → needle (see fig. 19). Needles, sector plates and dendritic crystals grow only if the supersaturation exceeds values which correspond roughly to the air being saturated with respect to liquid water.

The effect of suddenly changing the temperature and supersaturation on the growth form of a particular crystal may be observed simply by raising or lowering the fibre in the chamber. Whenever a crystal is thus transferred into a new environment, the continued growth assumes a new habit characteristic of the new conditions. Plate xvii(a) shows that when needles grown at about -5 °C were lowered in the chamber to where the temperature was about -14 °C, stars grew on their ends, and when similar needles were raised to about -2 °C, they gave way to hexagonal plates as shown in pl. xvii(b). In a similar manner it has been possible to produce combination forms of all the basic crystal types. Such radical changes in crystal shape could not be produced by varying the supersaturation at constant temperature, but in some cases were produced by only one degree change in temperature at constant supersaturation.

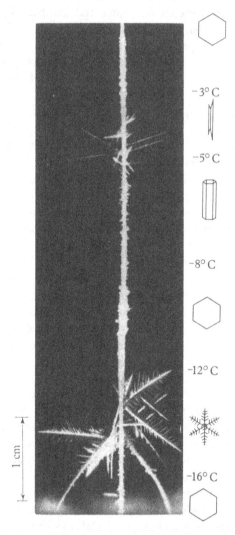

Plate xv. Ice crystals of differing shapes growing on a filament suspended in a diffusion chamber with controlled temperature gradient. The crystals take characteristic forms at different temperatures as indicated along the right edge of the photograph. Reading from the top, the symbols represent: thin hexagonal plates, needles, hollow prismatic columns, hexagonal plates, branched fern-like crystals (or dendrites), and hexagonal plates. At temperatures below − 25 °C, prisms appear again.

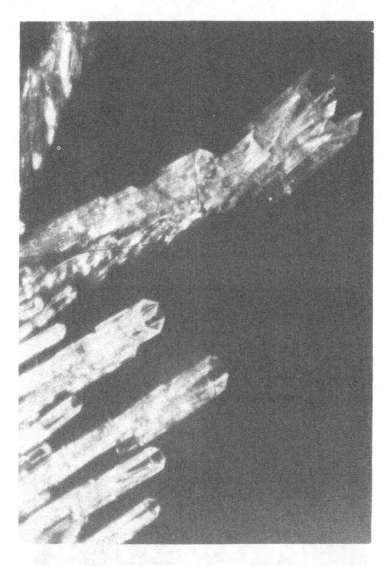

Plate XVI. Hollow prismatic columns grown from the vapour of heavy water.

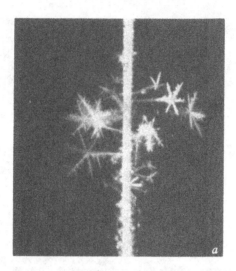

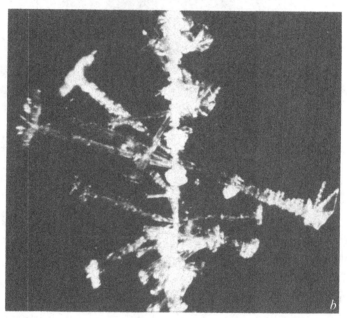

Plate XVII. Crystal hybrids showing how the form is dictated by temperature. Needles grown at −5 °C developed stars on the ends when shifted to a temperature of −14 °C (*a*); plates when shifted to −2 °C (*b*).

Experiments in which the crystals were grown in atmospheres of hydrogen, helium, carbon dioxide, and air at reduced pressure showed that their habit is not affected by the nature and pressure of the surrounding gas although, again, their growth rates are affected. On the other hand, the introduction of small quantities of certain organic vapours profoundly influences the crystal habit. These are strongly adsorbed on the crystal surfaces and may prevent certain faces from growing altogether.

The exact nature of the growth mechanism by which only a degree or two variation in temperature can completely change the crystal shape and which, furthermore, allows five such changes to occur in a temperature range of only 25 ° C, is still something of a mystery.

The growth and evaporation rates of ice crystals are clearly of fundamental importance in the development of precipitation. Given the distribution of temperature and humidity within the fall path, these may be calculated in much the same way as for a water drop by approximating the crystal shape to that of an electrical conductor of known capacity, e.g., a sphere, circular disk or ellipsoid. Measurements made on small spheres and crystals in the laboratory and in natural supercooled fog seeded with silver iodide are mostly in reasonable agreement with the theory but more accurate measurements over longer periods of growth in carefully controlled conditions are required.

The aggregation of snow crystals

Since the melting of even the largest individual snow crystals would produce only small drizzle drops, the aggregation of snow crystals to form snowflakes is an important step in the formation of precipitation from layer clouds. Clusters composed of a few individual crystals may arise as the result of hydrodynamic attraction between neighbouring crystals whose wakes extend for some tens of crystal diameters; alternatively, supercooled droplets caught by a crystal may subsequently freeze and grow into crystals. These small aggregates may then continue to grow by collecting smaller crystals by virtue of their greater falling speed. However, the aggregation of many snow crystals to form large snowflakes occurs only at air temperatures just below 0 °C and the question arises as to how the crystals stick together. It appears that adhesion is effected partly by the interlocking of crystals, by sintering of ice at their points of contact, and by the deposition of water vapour and the freezing of supercooled droplets acting as a cement. Sintering of ice occurs mainly by the evaporation of material from the convex facets of the crystal surface and its condensation into the concavities, the rate of sintering increasing rapidly with decreasing particle size so that spheres with radii of order 10 μ become firmly bonded together within only a few seconds. However, the fact that large snowflakes are found only near the 0 °C level and in parts of the cloud containing supercooled droplets suggests that riming and vapour deposition are more effective than sintering in causing the adhesion of crystal aggregates although more work is necessary to establish this firmly. There is also radar evidence to suggest that aggregation proceeds quite rapidly between partially melted snowflakes in the first hundred metres or so below the 0 °C level.

Experiments

1. Observations on the supercooling and freezing of water drops

The reader may observe the supercooling of water drops and check for himself that the freezing temperatures of, say, 100 drops are distributed in the manner of fig. 14, with rather simple equipment.

A refrigerating chamber may be constructed from two metal cans placed one inside the other with the space in between them filled with a three-to-one mixture of finely crushed ice and salt. The whole assembly should be lagged with thick felt or other suitable heat insulating material, and the top covered with a thick board until the temperature inside the inner can has fallen to −15 °C. The temperature is conveniently measured with a small thermocouple, the other junction being placed in a thermos flask containing a mixture of ice and distilled water.

The supercooling of a single drop of water may be demonstrated by suspending it from the end of the thermocouple and placing it inside the cold box. As the drop cools down the temperature indicated by the thermocouple will fall and eventually drop below 0 °C. The drop is now supercooled; check that it is still liquid. Allow the drop to cool further until it freezes, keeping a close, continuous watch on the temperature which will rise suddenly to 0 °C when freezing begins and then fall again as the *frozen* drop gradually cools to the air temperature.

To observe the supercooling of a group of drops, one needs a small metal dish partly filled with carbon tetrachloride with a layer of liquid paraffin floating on top. Water drops of about 2 mm dia., which may be introduced into these liquids from a hypodermic syringe, become suspended at the interface of these two liquids where the temperature can be measured by a thermocouple. When the dish is placed in the cold chamber, the drops will become supercooled and their freezing may be detected by the fact that the surfaces become rough and opaque. Also, the frozen drops, which contain air bubbles, often rise to the surface of the paraffin. The freezing temperatures of the drops may be recorded and their distribution plotted. Unless the water is highly impure, many of the drops may supercool below -15 °C, in which case another coolant, such as petrol chilled by solid carbon-dioxide (dry ice), should be used instead of ice and salt.

2. Detection and counting of ice nuclei

The cold chamber just described may be easily utilized for the detection and counting of ice nuclei. The interior of the inner metal can should be lined with black cloth or velvet tacked to a wooden frame and spaced $\frac{1}{4}$ in. from the walls. Breathing into the chamber will produce a dense white fog which will show clearly in a parallel beam of light from a very strong torch or projector lamp traversing the chamber diagonally. The presence of ice nuclei in the air will be revealed by the appearance of small ice crystals twinkling against the dull, grey background of the supercooled fog. A rough determination of their concentration may be made by estimating either the total number appearing in the known volume of the beam or their average distance apart. An average spacing of 1 cm will correspond to a concentration of 1 nucleus/cm³, while a spacing of 10 cm will imply a concentration of 1/l. Low concentrations of nuclei can be detected most conveniently by the supercooled sugar solution method described on page 64.

Progressive cooling of the chamber allows the nucleus concentrations to be determined at successively lower temperatures. It is important, however, to ensure that the splintering of frost crystals formed on the walls of the chamber do not give spuriously high counts; this is the main reason for the cloth or velvet lining which also provides a dark background against which to view the white fog and crystals. Frost formation may best be inhibited by coating the walls with a thin film of glycerol.

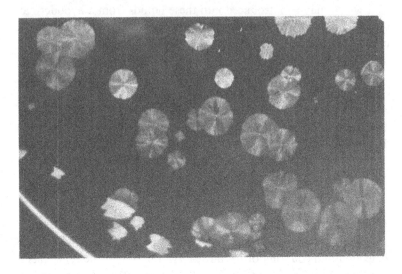

Plate xviii. A supercooled soap film used as a simple detector of ice nuclei. The ice nuclei nucleate the supercooled water in the film and initiate the growth of large crystals which are easily seen and counted.

3. A very simple ice-nucleus detector

A supercooled soap film or bubble makes a very simple and convenient ice-nucleus detector excellent for demonstration purposes. A very stable, long-lasting soap film stretched across a metal ring several inches in diameter is made by dipping the ring, soldered to a straight length of wire to serve as a handle, into a half-and-half mixture of the liquid detergent 'Stergene' and water. When the film is lowered into a cold chamber, submicroscopic ice nuclei in the air fall on to the film and are revealed by the appearance of large thin ice crystals which are easily seen and counted (see pl. xviii).

4. Making replicas of snow crystals

A solution of Formvar (polyvinyl formal) in ethylene dichloride is kept at a temperature not colder than $-5\,°C$, a clean glass microscope slide is immersed in the solution for about 30 s and then exposed horizontally to the falling snow crystals. A captured crystal becomes submerged in the solution after which the slide should be kept at a subfreezing temperature for a few minutes until the solvent evaporates and leaves the crystal encased in a thin, but tough, plastic shell. The slide may now be exposed to room temperature, when the crystal will melt, the water diffusing out of the plastic membrane as it evaporates, to leave a casting which retains, in microscopic detail, the surface structure of the original crystal.

A little practice and care is required to get the best results. It is particularly important to use solution of the right strength. If it is too viscous, small crystals will not become submerged and merely make a crater on the surface; if it is too thin, it will run off the slide and not cover a large crystal. Good replicas of natural snow crystals may be obtained with a solution containing one to three per cent by weight of Formvar, but to fix large snowflakes it is advisable to cover them with a drop of solution from a glass rod or an eye-dropper. The replicas may be stored and photographed at leisure.

5

SNOW, RAIN AND HAIL

Introduction

Inside all clouds the processes of condensation and aggregation produce particles of larger and larger size. Precipitation occurs when some of these particles reach such a size that they fall out of the clouds and the upcurrents which sustain them. If the particles are able to survive the evaporation which they experience while falling through the unsaturated air below the clouds, the precipitation reaches the ground; otherwise it may be noticed as fall-streaks or virga, which hang for some distance below the cloud base.

The size of precipitation particles is therefore partly determined by the strength of the updraught producing the cloud and by the humidity in the subcloud layer. Widespread layer clouds are associated with upcurrents usually less than 50 cm/s, so that droplets with radii exceeding 80μ can fall out of them and approach the ground. The distance which a drop can fall through unsaturated air before completely evaporating increases rapidly with increasing drop size; in an atmosphere of 90 % humidity, a droplet of radius 10μ will fall 3 cm before evaporating, while drops of 100μ and 1 mm would fall 150 m and 40 km respectively. Since the bases of dense clouds generally lie at least a few hundred metres above the ground, a radius of 100μ may be regarded as a lower limit for the size of the precipitation elements.

Forms of precipitation

Precipitation composed entirely of drops a little larger than this commonly falls in damp weather from low, shallow, layer clouds and is called *drizzle*. *Rain*, consisting of larger drops having radii up to about 2 mm, is produced by layer clouds some kilometres deep such as are associated with fronts and depressions. The heaviest rains composed of even larger drops fall from cumulus clouds whose depth may reach 10 km, and which contain powerful upcurrents of several metres per second. Precipitation from these clouds is sporadic, for their horizontal extent is only a few kilometres and the active life of individuals is less than one hour. These rains are described as *showers*.

The upper parts of the cloud systems which extend well above the 0 °C level contain ice particles. In stratiform clouds, these will usually be snow crystals of the type described in the last chapter. These join together to form snowflakes which fall at speeds up to about 1 m/s and either melt on crossing the 0 °C level or, in cold weather, reach the ground as *snow*. In shower clouds, the ice particles are usually in the form of hail pellets which may often melt before reaching the ground. However, very deep, vigorous clouds may produce very large hailstones which, in hot climates, are occasionally as large as oranges.

The occurrence of precipitation is strongly controlled by the motion of the cloud air. We have seen in previous chapters how the air motion during the formation of the cloud, in combination with the abundance and properties of those aerosols* it contains and which act as condensation and freezing nuclei, determines the concentration, initial size distribution, and nature of the cloud elements. As soon as these have formed, the processes of condensation and aggregation allow some particles to grow preferentially and the production of precipitation elements is now under way. Because the air motion governs the dimensions, water content, and duration of the cloud, it controls not only the rates of these processes, but the period during which they operate and thus the maximum size which the largest cloud particles can attain. If precipitation should result, it is again the air motion which determines the maximum rate of precipitation and its duration. The most important factors are therefore the air motion and its aerosol content, and from specifications of these it should be possible, in principle, to calculate the course of cloud and precipitation development. Unfortunately, we understand rather little about the motion of the air, and the laws governing the growth and aggregation of cloud particles are not yet firmly established. Moreover, the large variations in the concentration and properties of atmospheric aerosols, and the great complexity of atmospheric motions (as revealed by the great variety in the pattern of cloud development), add to the difficulty of constructing a detailed, general theory of precipitation development. However, considerable progress has been made with calculations, based on simple models of the air

---

* By aerosol we mean airborne particles.

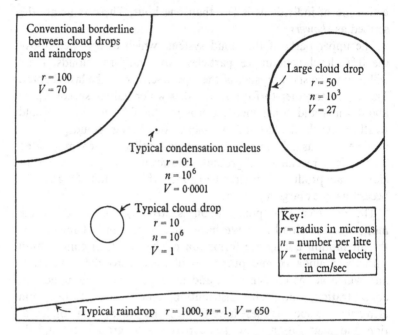

Fig. 20. Comparative sizes, concentrations, and terminal falling velocities of some particles involved in condensation and precipitation processes. (From McDonald, *Advances in Geophysics*, **5** (1958), 223.)

motion, of the growth of individual raindrops and hailstones and, as we shall see, these give a reasonable qualitative explanation of the formation of precipitation in different kinds of cloud.

The role of condensation

In droplet clouds the condensation process alone is ineffective in producing rain. The size which droplets attain by condensation is mainly determined by their concentration and the quantity of water vapour made available by the ascent of the cloud air. In the lower and middle troposphere, where the droplet concentrations established during the formation of clouds are usually of order $100/cm^3$, and the quantity of water vapour condensed is limited to about $7 \, g/m^3$, the *average* radius of cloud droplets cannot exceed 30μ. Droplet concentrations arising during condensation in very clean air at high levels may be of order $10/cm^3$, but the amount of vapour to be condensed is also greatly reduced, and so the conclusion that condensation alone is unlikely to produce droplets of *average* radius greater than 30μ is not affected. Droplets of this size evaporate completely after falling a few metres in unsaturated air, and so are incapable of reaching the ground as precipitation. Condensation on giant hygroscopic nuclei may produce *a few* larger droplets but these will attain radii of 100μ by condensation only after several hours. In the layer-type clouds which last this long the vertical air velocities are considerably less than 1 m/s, so that the droplets would fall out of them before reaching even this size. On the other hand, the orographic and cumulus clouds which contain updraughts capable of sustaining these drops do not provide more than an hour for droplet growth. Consequently condensation by itself is not capable of causing precipitation from droplet clouds.

Our central problem is then to explain how precipitation particles, at least one million times greater in volume than typical cloud droplets, are produced within about one hour of the cloudy air first becoming saturated. The comparative sizes, concentrations, and terminal falling speeds of the main types of particles occurring in clouds are drawn to scale in fig. 20.

Precipitation from layer clouds composed of droplets

In wholly liquid clouds the production of drizzle and raindrops is effected by collisions and coalescences between droplets of differing sizes. It is usually assumed that, among the main population of cloud droplets, there is usually a few larger droplets of radius 20–30μ. These may arise either on giant hygroscopic nuclei or by prolonged condensation on a small fraction of droplets which have remained, by chance, for a long time within the cloud without having been carried to the cloud edges by eddying motions and there exposed to evaporation.

If it is assumed that there are only feeble ascending motions (of less than 10 cm/s) within a well-stirred, low-level layer cloud it can be shown that such a large droplet can reach the size of a drizzle drop ($r \simeq 150\,\mu$) in falling from the top of a cloud about 1 km thick and having an average liquid-water content of only a few tenths of a gramme per m^3. If more substantial ascending motions (say, 50 cm/s) are assumed in the cloud, the thickness required for drizzle-drop growth is now about 500 m and small raindrops of radius $350\,\mu$ could form in a cloud 1 km deep.

These results accord with the observation that drizzle readily falls from low stratus clouds several hundred metres thick, particularly near hills and coasts (where the thickness may locally increase and where a more pronounced vertical motion may be introduced by orography), and that rain may fall when the clouds are more than about 1 km thick. Apparently similar clouds occur frequently without producing any precipitation; it may be supposed that these are clouds which have formed only very recently, or which contain unusually little condensed water or unusually small droplets, or in which the large droplets required to initiate the coalescence process did not form. Layer clouds of more than the minimum thickness of about 1 km rarely occur well above the ground except in storm regions, when the great thickness and extent of the layers, and persistent vertical motions of 10 cm/s or more, make the production of widespread rains practically inevitable.

Precipitation from layer clouds containing ice crystals

At temperatures only a little below 0 °C the concentrations of ice crystals which form in clouds are extremely low, and the difference in the saturation vapour pressures over ice and liquid water is small, so that the growth rate of a crystal appearing in a supercooled cloud is little different from that of the cloud droplets. At temperatures below -10 °C, however, crystals may appear in concentrations which are significant for the production of appreciable precipitation, and yet low enough to allow a moderate vertical air motion (greater than 10 cm/s) to keep the air nearly saturated with respect to liquid water. Since, at -10 °C, air saturated with respect to liquid water is supersaturated relative to ice by 10 %* and, at -20 °C, by 21 %, ice crystals will grow much more rapidly at these temperatures than will the cloud droplets for which the supersaturation will usually be less than 1 %. The rate of growth of an ice crystal may be represented by an equation very similar to (3·6):

$$\frac{dm}{dt} = 4\pi C(S-1) \Big/ \left[\frac{L_s}{KT}\left(\frac{L_s M}{RT} - 1\right) + \frac{RT}{DMp_\infty}\right], \qquad (5\cdot1)$$

* The saturation vapour pressure over a plane surface of pure liquid water, p_w, at -10 °C, is 2·86 mb; that over a plane surface of ice, p_i, is 2·60 mb. Thus

$$p_w/p_i = 2\cdot86/2\cdot60 = 1\cdot1 = 110 \%.$$

where S is the saturation ratio of the air, C is factor representing the shape of the crystal and L_s is the latent heat of sublimation of ice. For a sphere $C = r$, the radius of the sphere; and for a plate crystal, which may be represented by a thin disc, $C = 2r/\pi$.

Application of (5·1) indicates that when the crystals form in deep layers of cloud they may attain a dimension (plate-diameter or prism-length) of 1–2 mm after falling up to 2 km relative to the air during the course of an hour. A crystal of this size, on melting, would produce a drop of radius about 250μ. Even under the most favourable conditions growth would need to be prolonged for several hours, and over a fall-path of several kilometres, to produce a drop of twice this radius. In general, therefore, the growth of individual crystals by condensation is halted by their arrival at the melting level before their melted size has exceeded that of drizzle drops. However, in the neighbourhood of the melting layer the individual crystals are often aggregated into snowflakes, 10–100 or even more composing a single flake (see pl. xiv(c)). This aggregation of crystals is thought to occur as a result of collisions due to differing settling speeds or relative horizontal motions during a fluttering fall. After the snowflakes have melted the resulting drops may grow considerably by collision with cloud droplets at temperatures above 0 °C before they finally reach the ground as rain. This mechanism was first suggested by A. Wegener in 1911 and largely developed by T. Bergeron in 1933.

Above the melting level collisions with supercooled droplets may sometimes contribute to the growth of the ice crystals which now become covered with tiny frozen droplets known as rime. However, extensive areas of supercooled water in the snow region are rather unusual; they tend to be localized in regions where stronger vertical motions are produced by orographic or convective disturbances and, in general, there are only snow crystals well above the melting level. It follows that the freezing of droplets cannot be the source of crystals which is needed to replace those falling out of layer clouds as snow, and it is thought that the splintering of the large, fragile, fern-like snow crystals makes this provision.

The exploration of layer clouds by radar and aircraft

Radar, particularly when used in conjunction with instrumented aircraft and supplemented by careful visual observations of the cloud, has proved a very valuable research tool to the cloud physicist.* The fact that clouds may reflect back sufficient of the energy of the radar beam to be detectable by a sensitive radar receiver many miles distant enables him to study these clouds in the comfort of the laboratory. A small fraction of the incident energy is scattered back to the receiver-transmitter by the particles that compose the cloud, the actual fraction being determined by the size and concentration of the particles and the radar wavelength. This reflected signal is detected by the radar receiver, greatly amplified, and displayed on the screen of a cathode-ray tube to produce the radar 'echo' from the cloud.

Most present-day radars work on wavelengths of either 3 cm or 10 cm; usually these are capable of detecting a cloud only if it contains large particles, that is raindrops, snowflakes, ice pellets or hailstones. The echoes returned from the non-precipitating clouds are usually much too weak to be detected above the noise in the radar receiver. Radar may be used to provide information on the location of storms in space, the general shape, extent and movement of the precipitation areas, and the distribution and growth of the precipitation particles within the storm.

The radar echoes received from the widespread, continuous precipitation normally associated with warm fronts and depressions are characterized by the appearance of a particularly strong signal in the vicinity of the 0 °C level; this is generally called the *melting band* and is shown in pl. XIX(a) and (b). This separates a relatively steady echo obtained from the snow above the melting region and a rapidly fluctuating echo from the rain below. Above the 0 °C level the scattering particles are ice crystals and snowflakes, growth and aggregation of which cause the strength of the radar echo to increase with decreasing height. As the snowflakes fall through the 0 °C level they start to melt. Since a water particle reflects the radar waves more strongly than an ice particle of the same volume, the melting of the snowflakes is accompanied by at least a fivefold increase in echo intensity, which may be enhanced further

* For a detailed account of radar studies of clouds see Mason, *Physics of Clouds*, chap. 8.

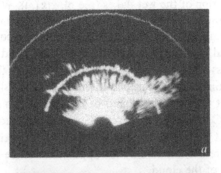

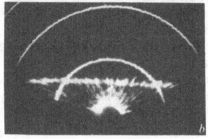

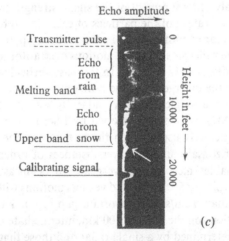

Plate XIX. (a) Radar echoes from warm-front precipitation showing the melting band and echoes from snow above and rain beneath it. (b) The sensitivity of the radar has been reduced so that the weaker echo from the snow has disappeared and shows the stronger echo from the melting band more clearly. (c) Radar echo from warm-front precipitation obtained on a vertically-pointing radar as a plot of echo amplitude against height. The picture shows a well-defined melting band at about 7000 ft. (Photograph by I. C. Browne.)

by the aggregation of the wet flakes (see pl. xix (c)). When the snow-flakes have completely melted, they collapse to form nearly spherical raindrops having considerably larger fall-speeds and correspondingly smaller spatial concentrations than the flakes, and also smaller scattering cross-sections. On both counts, just below the level where complete melting occurs, the echo intensity shows a sharp decrease, with the result that a narrow band of enhanced echo shows up just below the 0 °C level. The presence of the melting band is a certain indication that the rain is being produced by the melting of snow-flakes and therefore initiated by the appearance of ice crystals in the upper regions of the cloud.

By far the most important technique currently available for obtaining *simultaneous* information on the size and growth of hydro-meteors and the air motions within a cloud, is pulsed Doppler radar. Indeed this is the only available technique for studying the correlation between particle growth and the air motions. When pointed vertically in precipitation falling from stratiform clouds in which both the mean vertical velocity and the turbulence may be neglected, the spectrum of Doppler frequencies can be translated to give the falling speeds and hence the size distribution of the particles at all levels simultaneously. Measurement of the signal strength then allows the number concentration of the particles of each size to be determined.

Doppler radar cannot be used in the vertically pointing mode for measuring the widespread vertical motions of stratiform clouds which are of the order of only 10 cm/s. In this case vertical velocity is best derived from the divergence of the horizontal wind field that can be obtained from a single Doppler radar by means of the conical scanning (VAD) technique developed by Lhermitte and Atlas, and which is particularly useful when the airflow and the precipitation is almost horizontally uniform over distances of order 10 km. This technique enables meso-scale divergence to be measured with an accuracy of typically 5×10^{-5}/s, and vertical motions with an accuracy often better than 5 cm/s. An important gap in our knowledge con-cerns air motions on the scale of 100 km, intermediate between those that can be determined by a single radar and those that can be com-puted from routine rawin-sonde ascents, and which may account for much of the meso-scale structure of frontal precipitation. The UK Meteorological Office is measuring vertical motions on this scale from the divergence of the horizontal wind field, the latter being determined over areas of $(100 \text{ km})^2$ by releasing an array of self-opening radar targets from an aircraft at a height of about 6 km and following them with precision tracking radars as they fall at about 6 m/s. The position of each target is located very second, allowing horizontal winds averaged over height intervals of 500 m to be determined to within ± 20 cm/s and, with targets spaced at 25 km, vertical velocities to be determined to within ± 2 cm/s.

Very valuable information on the structure and evolution of extensive layer-cloud systems, usually associated with vigorous depressions, have been obtained in the United States by the simultaneous use of ground radar and an instrumented aircraft. The aircraft was equipped to measure the temperature, pressure and humidity of the air, the liquid-water content of clouds, and carried a rate-of-icing meter, a device for collecting ice crystals, and an accelerometer for indicating the turbulence.

In a typical intense cyclonic cloud system, the precipitation was heaviest near the centre of the storm where there were usually localized regions of strong vertical motions producing clouds of high liquid-water content in which soft hail and raindrops grew rapidly by accretion. The intensities of the vertical motion and of the precipitation fell off inversely as the distance from the storm centre and, in the outer regions, weak, widespread, uniform ascending motions gave rise to uniform layer-type radar echoes and precipitation of low intensity. But these cyclonic cloud systems are reluctant to conform to a 'text-book' pattern; one such, which was explored with radar and aircraft over a period of nearly 4 h, is depicted in figs. 21–24.

Figure 21 shows the pressure and wind distribution and the position of the surface fronts at the middle of the period, together with the distribution of rain which fell within the previous hour. Low clouds covered most of the storm area, but instead of the rain being most intense in the centre of the depression, it actually ceased along a line through the centre and perpendicular to its direction of travel and no precipitation occurred along the cold front. The heaviest rain occurred well ahead of the centre, about 200 miles to the north.

In the south-western quadrant of the depression, which was moving N.N.E., there were shallow altocumulus layers at 17000 ft and 11000 ft, and another at 8000 ft which merged with the lower-level cloud layers. Between 30000 and 17000 ft, the air was descending and was relatively dry. In the south-eastern quadrant there was clear, dry, descending air above 8000 ft and a broken altocumulus deck just below. The northern quadrants were regions of ascending

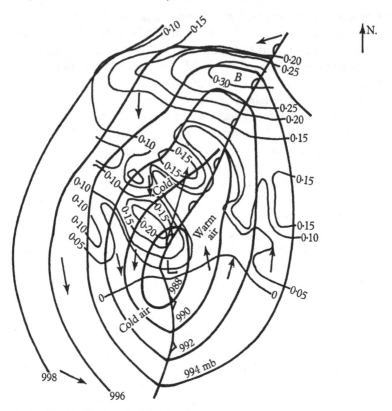

Fig. 21. The distribution of rainfall in a depression. The thick lines are isobars drawn at intervals of 2 mb showing the distribution of surface pressure. The thinner lines show the surface distribution of rainfall during the preceding hour in intervals of 0·05 in. The arrows show the directions of the surface winds. (After Cunningham, *M.I.T. Weather Radar Research*, Report no. 18, 1952.)

air; the vertical motion was rather weak near the centre of the depression but, farther north, there was active convection at middle levels between 11 000 and 17 000 ft surmounted by extensive canopies of cirrus and cirrostratus. It was here, the region *B* in fig. 21, that the heaviest rains, amounting to 0·3 in. in 1 h, occurred. Further ahead, some 400 miles from the centre, the ascending motions occurred at high levels to form cirrostratus and altostratus decks whose precipitation evaporated before reaching the low cloud layers.

Figure 23 shows a vertical cross-section of the cloud system along the flight path marked *DD* in fig. 22. An aircraft ascent along the vertical marked *A* revealed the presence of a uniform sheet

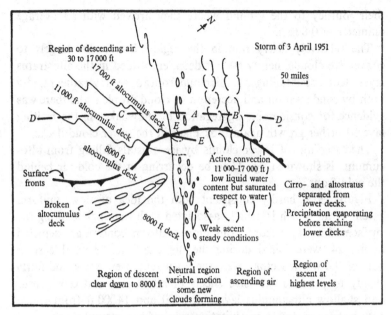

Fig. 22. Sketch map of the storm looking down from above showing the distribution of cloud relative to the surface fronts and the flight paths of the aircraft. (After Cunningham, *loc. cit.*)

of cirrostratus at 25000 ft through which the sun was dimly visible surrounded by a faint halo. Hexagonal ice plates and prismatic crystals, of diameter about 200μ, were present in roughly equal concentrations. Between 13500 and 14000 ft, where the temperature was about $-10\ °C$, there was a narrow band of altocumulus containing liquid water in quantities of about $0.05\ g/m^3$. Below 14000 ft there were dendritic ice crystals and small snowflakes, but in the layer between 13000 ft and the melting layer at 9000 ft, no liquid water was detected. The snow melted rapidly below 9000 ft to give a very shallow radar melting band; the absence of large wet flakes indicated that there was little aggregation. Just below the base of the melting layer, a thin stratocumulus (or nimbostratus) deck, broken in places, contained liquid water in concentrations of 0.3–$0.4\ g/m^3$. Clouds were absent between 7700 and 5000 ft but, below this, there was a dense layer formed by mixing, convergence and lifting of the air along the frontal zone. The average water content of this sheet was $0.8\ g/m^3$, so that the melted snowflakes grew appreciably by sweeping up the cloud droplets in this layer on the last stage of their journey to the ground where they arrived with an average diameter of 0·8 mm.

The relatively heavy rain in the region *B* was due largely to convective clouds, nearly 6000 ft deep, embedded in the altostratus layer. Ice crystals falling into this cloud from above grew rather rapidly both by condensation and accretion of cloud droplets and there was evidence for considerable aggregation of snowflakes in the melting layer. Further growth again took place in the lower cloud decks.

The 'seeding' of lower clouds by ice crystals falling from alto-cumulus is shown in fig. 23 to be occurring in the cold air behind the cold front.

Figure 23 is hardly consistent with the idea of slow, uniform ascent of air which is usually associated with the upglide along the surface of a warm front. But more uniform conditions, depicted in fig. 24, were found during the flight *EE* in the north-eastern part of the warm sector. Here the upcurrents were weak and fairly steady to produce extensive sheets of altostratus and cirrostratus and shallow altocumulus layers at 8000 and 14000 ft from which continuous, moderate precipitation reached the ground.

Precipitation from shower clouds composed of droplets

Showery precipitation from cumuliform clouds is generally of greater intensity and shorter duration than that from layer clouds, the elements usually being of larger mass. The particles which fall from vigorous shower clouds and thunderstorms are usually large raindrops and hailstones, while snow and small pellets of soft hail are usually associated with weak or decaying clouds. The large vertical displacements and correspondingly high liquid-water content of the air which forms the upper parts of tall clouds favour rapid growth of the precipitation elements by the accretion of cloud water. Moreover, the characteristically strong updraughts tend to keep the growing precipitation particles suspended in these regions of high cloud-water content. On the other hand, the strength of the updraughts is such that the individual cloud masses may reach their level of maximum development and be exposed to mixing with the clear air and evaporate before the accretion processes have the opportunity to produce precipitation particles. Evidently it is these features—the rate of the accretion processes (governed largely by the concentration of cloud water), the thickness of the

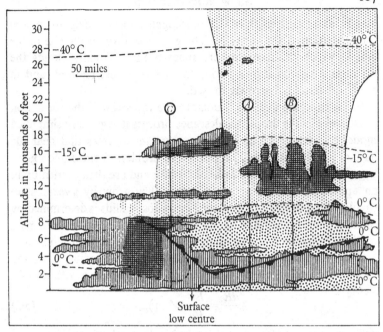

Fig. 23. Vertical cross-section, based on aircraft observations, through an extensive cyclonic layer-cloud system. The flight path is that marked *DD* in fig. 22. The close dotting represents ice crystals or snow, the vertical hatching water cloud, the cross-hatching areas of both supercooled water and snow, and the open dotting rain. (After Cunningham, *loc. cit.*)

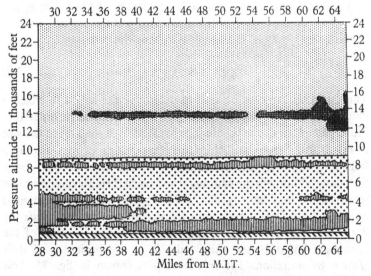

Fig. 24. Vertical cross-section through the cloud system along the flight path *EE* in the warm sector of the depression. The key to the shading is the same as for fig. 23. (After Cunningham, *loc. cit.*)

clouds, and the strength of the updraughts which principally govern the formation of shower rains. The great variation in the size of the clouds which produce showers suggests that in many clouds the favourable combination of conditions is only just realized, and in many others only narrowly missed.

E. G. Bowen and F. H. Ludlam have investigated the formation of showers by the droplet-coalescence process using a very simplified model of cumulus cloud structure. Ludlam assumes that near the base of the clouds a small proportion of droplets, formed upon giant hygroscopic nuclei, attain radii of 20–40μ and are then carried up in an updraught of constant speed when they grow by sweeping up smaller cloud droplets of uniform sizes whose radius is determined by their concentration and the cloud-water content. The latter is assumed to be that corresponding to adiabatic ascent of the air from cloud base.

The rate of growth of our larger droplet of radius R, falling speed V, sweeping up smaller droplets of fall-speed v with an efficiency E is

$$\frac{dR}{dt} = \frac{Ew}{4\rho_L}(V-v), \tag{5.2}$$

while the vertical motion of the growing drop relative to cloud base is described by

$$\frac{dz}{dt} = (U-V), \tag{5.3}$$

where U is the updraught velocity and w the cloud-water content.

It is now assumed that w is a function only of the height z, and $v \ll V$, so that the growth of the droplet from radius R_0 to R between levels z_0 and z may be evaluated by solving the equation

$$\int_{R_0}^{R} \frac{(U-V)}{EV} dR = \frac{1}{4\rho_L} \int_{z_0}^{z} w\,dz. \tag{5.4}$$

This equation can thus be used to determine the height to which a cloud containing a steady updraught U must extend in order that the largest of the droplets entering the cloud base shall reach the size of raindrops. Somewhat before this stage, the drops may acquire a fall-speed equal to, and subsequently greater than U. The drops then descend through the updraught and reach their maximum size on reaching the cloud base again. At this stage $z = z_0$, and the integral on the right of (5.4) is zero; thus the maximum size with which the drops will fall out of the cloud base is solely a function of the updraught speed U and their radii R_0 at the beginning of their growth by accretion. The relationship is shown in fig. 25. This diagram indicates that if the mean updraught in the cloud exceeds 3 or 4 m/s then raindrops may attain a diameter of 5 mm. Drops appreciably larger than this become distorted, oscillate as they fall, and are readily disrupted by small disturbances into several large fragments and many more small drops. These larger fragments

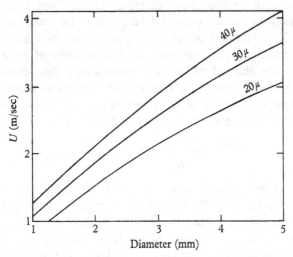

Fig. 25. Diameters of raindrops grown by coalescence from droplets of initial radii 20, 30, 40μ as a function of the velocity U of a steady updraught. (From Ludlam, *Quart. J. Roy. Met. Soc.* 77 (1951), 402.)

may grow rapidly by coalescence with cloud droplets and, in dense, vigorous clouds, may themselves reach the unstable size within 3 or 4 min. The possibility of a chain reaction of raindrop multiplication therefore arises, as first pointed out by Irving Langmuir. Figure 25 shows that, for a given updraught velocity, the largest raindrops result from smaller initial droplets as these are carried up further into the cloud and so have a larger fall-path. This does not necessarily mean that smaller droplets are more efficient in releasing rain, for although their growth is potentially greater it is accomplished only over a longer period which may extend beyond the life of the cloud.

So far, we have discussed only the *maximum* size which a raindrop *may* attain in a steady updraught and we have not considered the time and space required. Here the water content w of the cloud is of great importance. A low value of w might mean that the drop would be carried up to high levels before reaching a falling speed in excess of the updraught and there become frozen. Alternatively, it might be left behind by a dissolving cloud tower and not be large enough to survive evaporation before falling back into a newly rising cloud mass.

It is on this last possibility that Ludlam bases a criterion for shower development. He specifies that air carried to the cloud summits shall contain drops of radius greater than 150μ which, in falling at more than 1 m/s, will settle out of the weak updraughts within the cloud summits, survive evaporation during a fall of several hundred metres in clear air, and thus have a chance of being caught up in a new thermal where they can continue their growth. The *minimum* cloud depth, z_{min}, required for droplets of various initial radii to grow to 150μ while being carried from cloud base to summit in a steady updraught may be computed from (5·4):

$$\int_{R_0}^{R=150\mu} \frac{(U-V)}{EV} \, dR = \frac{1}{4\rho_L} \int_{z_0}^{z_{min}} w \, dz,$$

the results being shown in fig. 26. This suggests that, in a cloud having a warm base (20 °C) and feeble updraughts of 1–2 m/s, a shower may develop when the depth exceeds about $1\frac{1}{2}$ km; and if the base temperature exceeds about 10 °C, showers may form in clouds whose tops fail to reach the 0 °C level. The minimum cloud thickness required increases with the mean speed of the updraught and is also greater in colder clouds which contain less liquid water. The time required for the growth of a raindrop is found by adding a few minutes to the period obtained by dividing the minimum depth z_{min} by the updraught velocity U, and varies from about half an hour with $U \simeq 1$ m/s to about 20 min in the stronger updraughts. In order for a considerable shower to occur, the life of the entire cloud must exceed these periods by at least several minutes. The calculations also indicate that if the cloud-base temperature exceeds 10 °C, showers may fall from clouds whose summits do not reach the 0 °C isotherm. If the base temperature exceeds 20 °C, the shower-cloud tops are not supercooled even with updraughts of 8 m/s. On the other hand, in clouds with cold bases, the initial rapid growth of the ice particles by sublimation may allow them to release a shower before the coalescence process can get under way. In clouds of intermediate base temperatures, say -5 to $+5$ °C, either the one or the other process may dominate depending upon the updraught strength and the aerosol content, and indeed both may contribute to the formation of precipitation elements inside a cloud of more than

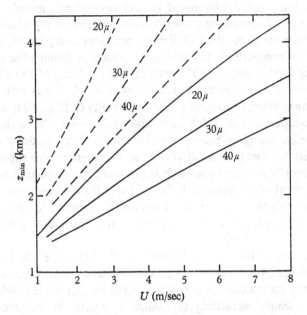

Fig. 26. The variation of minimum cloud depth z_{min} for shower development as a function of the updraught U, for initial droplet radii 20–40μ in the bases of clouds with base temperatures -5 °C (pecked lines) and 20 °C. (From Ludlam, *loc. cit.*)

critical size. In such clouds the complete description of shower formation and development must be very complicated, and in view of this and our rudimentary knowledge of the cloud dynamics, it is difficult to follow all its possible interactions.

In summary, the elementary calculations made by Ludlam probably describe quite well the relations between the updraught, cloud temperature and minimum depth of cloud necessary for shower production, and accord quite well with observations, but only because, on the one hand, they underestimate the growth rates of incipient precipitation elements by treating coalescence and accretion as continuous rather than as stochastic processes, and, on the other hand, they overestimate the liquid-water concentrations of the cloud by assuming the adiabatic values. Morever they beg the question of how droplets of radius 20–40 μ appear at cloud base in the first place. There is a need to revise these calculations using the stochastic approach to the growth process, using more recent and reliable computations of collection efficiency, and cloud models in which the updraught and liquid-water content vary with height in a realistic manner.

A specimen calculation based on the two-thermal model of a maritime cloud containing a low concentration (66 per mg of air) of salt nuclei none larger than 10^{-11} g and mentioned on page 57, may be cited for comparison. The first thermal ascends through the condensation level at 900 mb, 10 °C with an initial velocity of 1 m/s and excess temperature over the surroundings of 0·1 °C, mixes with the environment assumed to have a relative humidity of 85% and a lapse rate of temperature of 7 degC/km, and finally reaches a height of 330 m before losing its buoyancy and coming to rest. At this stage, the largest droplets have a radius of 18 μ. This thermal now subsides and practically evaporates before being caught up in a second, newly-rising thermal which mixes with the residue of the old one, achieving a maximum vertical velocity of 2·4 m/s and a maximum liquid-water content of 0·9 g/kg before coming to rest at a maximum height of 1·6 km above cloud base.

At this stage, after a total elapsed time of 35 min, the cloud tops contain drops of radius 100 μ in concentrations of 100/m³ and these would grow by coalescence to radius 150 μ in another 5 min at which size they would, according to Ludlam's criteria, be capable of initiating a shower.

The release of showers by the growth of ice particles

Although very little observational evidence is available, it is believed that ice crystals in concentrations significant for shower formation (of order 1/l.) form in the supercooled parts of cumulus when the temperature has fallen to between -10 and -15 °C, but that this threshold temperature varies considerably from day to day and from place to place. These variations are attributed to changes in the concentration and properties of atmospheric ice nuclei (see chapter 4). Very large variations are observed in the concentrations of crystals which appear when samples of outdoor air are introduced into a supercooled cloud formed in a cloud chamber, but it is not known exactly how these may be related to the concentrations which could occur when the air is involved in natural cloud formation and when crystals may be produced by the secondary processes described on page 68.

It has sometimes been assumed that shower formation occurs when cumulus summits reach a level at which ice crystals are present in the requisite concentration of about 100/m³, but clearly some additional development is required if the crystals are to reach a size which will ensure their continued growth inside the cloud into precipitation elements. By analogy with this criteria for shower production by the growth of water drops, Ludlam considers that ice particles must at least reach a size such that during their growth they have acquired falling speeds exceeding 1 m/s and can fall through several hundred metres of unsaturated air without significant evaporation.

The size at which these conditions are fulfilled depends upon the size and structure of the particle. Newly formed particles grow at first by deposition of vapour until they reach a diameter of about 100 μ beyond which these are able to collect the cloud droplets with increasing efficiency. At temperatures around $-15\,°C$ this first stage is accomplished in about 4 min and thereafter the capture and freezing of supercooled droplets leads to the formation of a roughly spherical aggregate containing many air enclosures. The density of such pellets of *soft hail* usually lies between 0·1 and 0·3 g/cm³. For such densities the particles must attain radii of 500 and 300μ respectively for the conditions listed above to be fulfilled. In a cold cloud with base temperature $-10\,°C$ and having an average liquid-water content of 0·4 g/m³, growth to such a size takes an additional 10 min, at the end of which the cloud summit will have reached a level where the temperature is as low as -20 or $-25\,°C$. The particle may subsequently grow, in the next 10–15 min to a radius exceeding 1 mm, fall out of the cloud and reach the ground as a pellet of soft hail. This type of precipitation commonly falls from cold shower clouds in the winter time. If the cloud-base temperature is well above 0 °C such particles will usually melt before reaching the ground. In warm, vigorous clouds containing high concentrations of liquid water, the ice particles are likely to develop into true hailstones. In these warm clouds, precipitation is also more likely to be released by the coalescence process. But, in many clouds both processes may contribute precipitation elements to a shower formed inside a cloud of more than minimum size, in which cases, the complete description of shower formation and development must be very complex.

Indeed, enough has been said to indicate that even a *qualitative* theory of shower production is complicated. The formation of a shower requires a favourable combination of large-scale and small-scale properties to ensure that the growth of precipitation elements is accomplished within the useful life of the cloud. The size, speed and distribution of cloud updraughts, the cloud-base temperature, the degree of mixing between the cloud and its surroundings, the distribution of humidity, temperature and wind in the environment, the concentration and nature of the atmospheric aerosol, are all factors which are important and which are subject to great natural variation. Perhaps it is not surprising that a *quantitative* theory of shower rains has not yet been attempted.

The formation of hail

Hailstorms occur most frequently in the continental interiors of middle latitudes and diminish towards the poles and the equator and over the sea. In cold climates the clouds are not sufficiently vigorous neither do they contain large enough concentrations of cloud water to produce large hailstones. Over the temperate oceans, in the absence of intense surface heating, the cumulonimbus do not become organized into the large cloud systems which we usually associate with large hail. In the tropics, the strong horizontal temperature gradients and strong vertical wind shears, which favour the development of damaging hailstorms, do not usually extend into the high troposphere. Although the occurrence of hail, particularly of large stones, is usually associated with thunderstorms, small hail may fall from clouds which do not reach thunderstorm proportions. Large hail usually falls from only rather localized regions of the storm and for a brief period. The areas of severe hail damage on the ground may vary in width from a few yards to several miles, commonly one mile. The duration of a hail fall may vary from as little as 10 s to as much as 30 or 40 min, a median value being, perhaps, 5 min.

Most true hailstones are either roughly spherical, conical, or discoidal in shape, the spherical forms being by far the most common, especially when the hail is small. Sometimes jagged irregular shapes with many external irregularities and protuberances are formed.

There have been few systematic observations of hailstone size and weight; observers tend to report the maximum and average sizes of hailstones, but not the smaller sizes, so that we have very little data on the size distributions of hailstones from individual storms. The distribution of sizes of the largest hailstones observed in storms occurring in the Denver area of Colorado during 1949–55 was:

Diameters of largest hailstones	No. of cases
Grain (< ¼ in.)	10
Currant (¼ in.)	122
Pea (½ in.)	282
Grape (¾ in.)	149
Walnut (1–1¼ in.)	38
Golf ball (1¾–2 in.)	26
Tennis ball (2½–3 in.)	4

Hailstones of at least walnut size fell on at least two days in each year.

The largest hailstones ever reported in the United States fell in Nebraska, the largest being the size of a grapefruit, 5·4 in. (13·8 cm.) in diameter and weighing 1½ lb.

The structure of hailstones

The size, shape and internal structure of large hailstones in terms of their modes of growth and fall in the cloud, topics which have exercised scientists for more than a century, have received satisfactory explanation only during the last decade. Hailstones may originate either as snow pellets, particles of small hail, or as particles of clear ice (perhaps frozen raindrops), and grow mainly by the deposition and freezing of supercooled cloud droplets. Within the complex structure of hailstones it is possible to distinguish three kinds of ice deposit, the structure and conditions for growth of which have been studied in laboratory experiments. If the impacting droplets are deeply supercooled they freeze rapidly as individual droplets to form a rime structure. The air spaces between the frozen droplets give the ice a low density usually somewhere in the range 0·1 to 0·6 g/cm³. The frozen droplets contain large numbers of tiny air bubbles which, because of their high scattering power for light, give the ice an opaque, white appearance. Soft hail pellets are largely composed of low-density rime and this is also found in the cores of some large hailstones. As pointed out by Schumann and later by Ludlam, there is a limit beyond which water deposited on an ice particle can no longer freeze completely because the latent heat of fusion cannot be dissipated sufficiently rapidly to the environment by forced convection and evaporation. When this critical limit is reached the droplets spread over the surface to form a continuous film before freezing and maintain the surface in a just-wet condition at 0 °C.

Under these conditions freezing proceeds slowly, releasing little air, so that the ice is clear and has a density close to 0·9 g/cm³. Such *compact ice* may also be formed on colder dry surfaces if the droplets impact at high speed and spread before freezing, in which case the ice may be opaque. Ludlam calculated the conditions under which a *smooth* ice sphere would become wet and deduced not only that large hailstones cannot grow in the dry state but that only that fraction of the collected water which actually freezes can contribute to the growth of the stone, the rest being shed in the wake of the hailstone. Shedding of the unfrozen water would certainly impose a severe limitation on the growth of large stones and this made it very difficult to account for the formation of the largest until List discovered that, in growing large wet aggregates of accreted ice in a wind tunnel, the excess water was not carried away but was retained by a skeletal framework of ice, the proportion of liquid water in this 'spongy' ice being determined by the heat exchange between the aggregate and the air-stream. This wet spongy ice, which has a density of 0·9–1·0 g/cm³, is often quite transparent, but sometimes contains sufficient concentrations of small air bubbles to give it a milky appearance.

The internal structure of large hailstones, as revealed by the examination of thin sections in ordinary and polarized light, consists of a growth centre or 'embryo', usually a few millimetres to 1 cm in diameter, followed by a number of alternate zones of transparent and relatively opaque ice—see pl. xx. The centres are most commonly opaque spheroids composed of rime ice, but clear spheroids are quite common and the largest hailstones quite often have conical embryos. Large hailstones usually have at least four well-defined layers of distinctly different crystal structure although as many as 10–20 layers may be distinguished by differences in opacity.

The crystalline structure of the layers, as revealed by the size, shape, spacing and orientation of the component crystals, is determined by several factors such as the concentration, size, and impact velocity of the collected supercooled droplets, the temperatures of the air and hailstone surface, and the detailed mechanisms of freezing. Laboratory experiments by Hallett indicate that, in the dry-growth regime, the supercooled droplets freezing on impact at temperatures above the critical value will tend to take the orientation of the underlying surface and produce a fabric of large crystals. This tendency will be even more pronounced if the droplets impact at high speed and spread out before freezing to form a more compact glaze ice. On

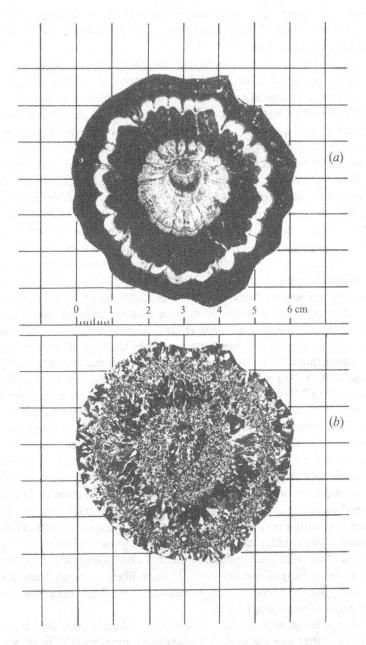

Plate xx. This section through the centre of a large lobed hailstone shows: (*a*) the bubble structure in reflected light; regions of clear ice appear black, and milky or opaque ice appears white; and (*b*) the crystal fabric photographed in transmitted polarized light. (Photograph by courtesy of Dr K. A. Browning.)

the other hand, at temperatures below $-15\,°C$, most droplets will freeze individually in random orientations and produce a low-density matrix of small crystals.

The appearance of an ice deposit, its opacity and whiteness, is due largely to the scattering of light by minute air bubbles trapped between, and especially inside, the frozen droplets. The opacity is determined by both the ambient temperature and the surface temperature of the deposit. The ambient temperature governs the concentration of air dissolved in the impinging droplets, the concentration increasing with decreasing temperature, and hence the quantity of air released during freezing. The temperature of the deposit determines the rate of freezing and hence whether the air bubbles become trapped in the ice rather than escape by migrating to the surface. Low temperatures favour the formation and retention of high concentrations of small air bubbles and hence the growth of opaque ice. The deposition and slow freezing of water at temperatures close to $0\,°C$ tends to produce clear ice, but the freezing of ice-water mixtures at low ambient temperatures often produces milky ice containing modest concentrations of air bubbles that are considerably larger than those appearing in opaque ice. Examination of the crystal structures that result from the complete freezing of spongy ice suggests that if the water is deposited at temperatures between $-5\,°C$ and $-20\,°C$, the initial freezing of the fraction required to raise the mixture to $0\,°C$ is accomplished by the growth of a three-dimensional array of dendrites. Later freezing of the water retained in this skeletal framework produces rather large single crystals, many with optic axes roughly tangential to the hailstone surface. When the deposited water is deeply supercooled, its dendritic growth is less orderly and freezing may start on the outside and spread inwards. The situation may be further complicated by the fracture of fragile dendrites and the capture of ice crystals, both of which give rise to new nucleation centres and the formation of a matrix of medium or small crystals with random orientation. The growth of large crystals of clear ice, with their optic axes oriented along the growth direction (i.e. the radius) of the hailstone, is more likely to result from the orderly growth of compact ice in the just-wet condition than from the freezing of wet spongy ice.

Large hailstones are often composed of a three-dimensional array of lobes that give the surface a convoluted appearance as in pl. xx. The lobes arise from surface protuberances which capture cloud droplets more efficiently than the adjacent depressed areas. Thus the lobes tend to grow into the airstream and, if the hailstone tumbles constantly during growth, the lobes develop into a three-dimensional array with radial symmetry. Bailey and Macklin have grown artificial lobed stones of up to 12 cm diameter in an icing tunnel, the lobes being most pronounced when the stone was growing near the wet limit and the accreted droplets were small. This implies that natural lobed hailstones grow in strong updraughts containing small cloud droplets and not in regions containing high concentrations of supercooled raindrops that would almost certainly produce spongy growth. The dissipation of latent heat from lobed stones is considerably higher than from the smooth spheres of the same size and it is the presence of these protuberances that allows the hailstones to grow to a large size without becoming excessively wet and spongy.

Theory of hailstone growth

The first really quantitative theory of hailstone growth was produced by Schumann in 1938, who assumed a hailstone to grow as a spherical particle by sweeping up all the supercooled water droplets lying in its fall path and, starting from a diameter of 0·5 cm, calculated its ultimate diameter in terms of the distance travelled, the liquid-water content, and the updraught velocity in the cloud. He concluded that, in order to produce hailstones of several centimetres in diameter, the clouds must extend to heights of several kilometres, and that they must contain either very high concentrations of liquid water or very strong updraughts. However, his calculations indicated that hailstones of 3 cm diameter could be produced in clouds containing only 4 g/m^3 of liquid water with updraughts of 10 m/s, which was only about one-half the terminal velocity of the stone. However, Schumann realized that these calculations were not complete in that he did not investigate the early, slow stages of growth and, moreover, assumed that all the water collected by the hailstone would freeze on its surface. In the second part of his paper, Schumann considered the heat balance of the hailstone and formulated relationships to determine, for given values of the ambient temperature, limiting values of the liquid-water content above which the water collected by the hailstone cannot all freeze immediately and some may be shed in the form of drops. His general treatment of the heat balance has been refined and improved by several writers in recent years. Ludlam, whose treatment is essentially correct for smooth ice spheres of

$d < 3$ cm, derived a criterion (see curve 3 of fig. 27) for the stone to become wet, *viz.* $RV > f_1(z)$, where R and V are respectively the radius and fall velocity of the stone and $f_1(z)$ is a function of height (and therefore temperature) above cloud base. If the liquid-water content of the cloud exceeds the critical value so that the stone becomes wet, Ludlam assumes that the excess accumulates as a liquid film or is shed and that in this wet régime transparent ice of density about 0·9 g/cm³ is deposited beneath the liquid film at a rate governed by the rate of heat transfer. The rate of growth of the stone in relation to its height above cloud base is then given by

$$(U - V)\left(\frac{R}{V}\right)^{\frac{1}{2}} \frac{dR}{dz} = f_2(z), \qquad (5\cdot5)$$

where U is the updraught velocity and $f_2(z)$ is again only a function of height or temperature. Figure 27 depicts the growth of two small drops, frozen at $-5\,^{\circ}\text{C}$ and $-10\,^{\circ}\text{C}$ respectively, in a persistent constant updraught of 20 m/s. They grow first as pellets of soft, dry hail of density 0·3 g/m³ (dotted lines) and then continue as wet particles but shedding all the water that cannot be frozen. Finally they attain radii at the 0 °C level of rather more than 1 cm, which is small by comparison with the diameters of 6–10 cm with which hailstones occasionally reach the ground.

But, as we have seen, the excess water is likely to be retained in a framework of spongy ice and so Ludlam's limitation is removed and there is the possibility of hailstones growing to diameters of several centimetres in the wet spongy state. Such hailstones are also markedly aspherical but it turns out that the conditions for the onset of wet growth on *smooth spheroids* are virtually independent of their eccentricity. The situation may, however, be quite different for stones having the lobed structure mentioned earlier since their heat transfer and drag coefficients are considerably greater than those for smooth spheres of the same mass. Bailey and Macklin find that the critical liquid-water concentrations required to cause wet growth are 1·2 to 3 times those computed for smooth spheres and so quite large lobed stones may grow in liquid-water contents of several g/m³ without becoming excessively wet and spongy. If a hailstone were to shed no water, then it could attain a radius of 3 cm in a water concentration of 3 g/m³ in about 10 min.

Large hailstones have fall speeds of 30 m/s or more, so any theory of their production requires that the cloud shall contain updraughts

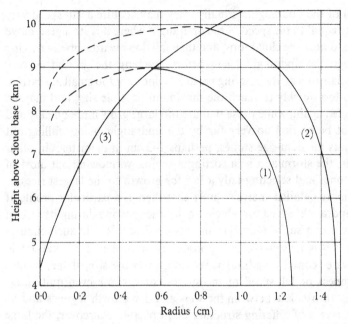

Fig. 27. Growth of hailstones in a cloud with base at 900 mb, 20 °C and following a saturated adiabatic. The curves (1) and (2) show the growth in a constant updraught of 20 m/s following the freezing of a cloud droplet at – 5 °C and – 10 °C respectively. Curve (3) shows the conditions under which a hailstone of density 0·9 g/cm³ becomes wet. (From Ludlam, *Nubila*, 1 (1958), 12.)

of comparable speed in order to keep the stones suspended for periods of about 10 min. In the early stages of growth, the fall-speed of the hailstone increases rather slowly so that a *strong steady* updraught will carry it up through the supercooled region before it can attain a large size. If the updraught were *intermittent* or *pulsating*, the hailstone might repeatedly fall from a high level and then be carried up again, but there is no evidence that strong organized updraughts are intermittent. Instead, Browning and Ludlam suggested that severe hailstorms may usually contain an almost steady strong updraught which, because of vertical wind shear through a considerable depth of the troposphere, is inclined to the vertical, so that particles falling from the high level outflow may re-enter the updraught at a lower level and make a further ascent.

If we now suppose that the updraught speed increases with height, then a small proportion of the re-entering stones that have a favourable size, may be lifted *slowly* by the updraught, growing at such a rate that the increase in their fall speed closely matches the increase in the speed of the updraught. Finally, they may acquire a fall speed very nearly equal to the speed of the updraught where it is strongest, move forward near the cloud tops, and then fall downwards, briefly passing through the updraught again before reaching the ground as very large hailstones. Re-entering particles which are too small are carried up to too quickly to reach the maximum possible size, and may be re-cycled again, while those that are too large grow more quickly and cannot be carried up very far by the updraught before falling out again as medium-size stones, perhaps 1–2 cm in diameter. In other words, the storm acts as a sorting machine, winnowing out most of the stones and selecting only a few for growth to the largest sizes.

A more detailed analysis shows that there is a particular profile of the updraught speed for which a hailstone growing during its second ascent has a surface temperature always just 0 °C. In such circumstances it is reasonable to suppose that rather small fluctuations in the water content, updraught velocity, or in the size, shape, surface roughness or fall-speed of the stones could cause even quite large stones to oscillate between the wet- and dry-growth régimes and so form layers of differing structure and opacity. Moreover, the large hailstones may seriously deplete the cloud water, and any appreciable change in their concentration brought about by, say, fluctuations in the updraught may, in turn, cause fluctuations in the water content and hence in the growth rate and structure of the stones.

Finally, one must mention an important recent analysis by Macklin, Merlivat and Stevenson which shows that from the isotopic composition of the various layers of a hailstone it is possible to infer the air temperatures, updraught speeds and cloud water concentration experienced by the stone during its growth and hence obtain a trajectory compatible with its internal air-bubble and crystalline structures. The analysis confirms a re-cycling process of the nature envisaged by Browning and Ludlam. The particular hailstone analysed ascended at least twice in the updraught. The zones of ascent in the cloud each gave rise to two growth layers in the stone, one clear and comprising quite large crystals, the other opaque and comprising relatively small crystals. The opaque layers were formed at air temperatures below − 20 °C. The sharp transition between the pairs of layers indicates that there was a zone of descent outside the main body of the updraught. The updraught profiles calculated for each of the ascents are consistent and are in agreement with that calculated on the basis of the parcel method.

It is important to verify by observation that the existence of a tilted updraught in severe storms is essential for the production of giant hail, because such a requirement would largely explain why such hail is mainly confined to the continental interiors of middle latitudes, where strong convection and wind shear tend to coincide, and why it is a rarity in the tropics.

6

RAINMAKING EXPERIMENTS

Historical introduction

There are on record a number of attempts by man to increase the rainfall by a variety of interesting methods such as the lighting of fires, the firing of cannon, the production of electric discharges by kites, and the spraying of liquid air and of dust from aeroplanes. The firing of guns and rockets and the ringing of church bells have long been practised in Switzerland, Italy and Austria as methods of preventing the formation of large, damaging hailstones. It is only since the Second World War, however, that methods based on a knowledge of the physical processes of rain formation have been used.

The demonstration in recent years that suitable clouds may be made to release their precipitation by introducing artificial nuclei, and the possibility of extending this technique to produce economically important increases of rainfall has created the greatest interest and a good deal of controversy. This is certainly an exciting development in atmospheric physics, and, whatever may be the ultimate consequences, it has given a new impetus to researches on cloud- and rain-forming processes.

Modern rainmaking experiments are based on three main assumptions:

(i) That either the presence of ice crystals in a supercooled cloud is necessary to release snow and rain by the Wegener–Bergeron process (see page 98), or the presence of comparatively large water droplets is necessary to initiate the coalescence process (see page 106).

(ii) That some clouds precipitate inefficiently or not at all, because these agents are naturally deficient.

(iii) That this deficiency can be remedied by seeding the clouds artificially with either solid carbon dioxide (dry ice) or silver iodide to produce ice crystals, or by introducing water droplets or large hygroscopic nuclei.

The possibility of producing rain and suppressing hail from supercooled clouds by the introduction of artificial ice nuclei was foreseen by the German cloud physicist Findeisen in 1938, but it was not until 1946 that a satisfactory method of supplying nuclei in the required large quantities was discovered. Earlier, in 1931, Veraart in Holland dropped dry ice, among other things, into supercooled clouds and must have produced such nuclei, but he was not aware of this. It is quite possible that Veraart produced slight amounts of rain on several occasions, but because of his too-sweeping claims, all his attempts were discredited.

Not until July 1946 when Schaefer, working at the General Electric Company in the United States, accidentally discovered that a tiny fragment of dry ice, when dropped into a cold chamber filled with supercooled cloud, resulted in the formation of several millions of ice crystals, was it feasible to put Findeisen's ideas to the test. Schaefer made the first field trial on 13 November 1946 when 3 lb of crushed dry ice were dropped along a line about 3 miles long into an altocumulus deck whose temperature was about $-20\,°C$. Dr Irving Langmuir, observing from the ground, saw snow fall from the seeded cloud for a distance of about 2000 ft before evaporating in the dry air. A number of such tests in the following months gave very similar results, that is large areas of supercooled stratiform cloud were converted into ice cloud. On some occasions, snow fell from the seeded part of the cloud to leave a clear lane.

A further important step in the history of cloud seeding was the discovery by Vonnegut in November 1946 that minute crystals of silver iodide, produced in the form of a smoke, acted as efficient ice-forming nuclei at temperatures below $-5\,°C$. The fact that enormous numbers of nuclei, about 10^{15} from 1 g of silver iodide, could be produced by vaporizing an acetone solution of silver iodide in a hot flame, immediately suggested the possibility of dispersing them in large quantities from the ground, since the minute particles could remain in the atmosphere for long periods until carried up by convection currents into the supercooled regions of the cloud.

The first clear-cut evidence that silver iodide smoke could modify natural, supercooled clouds was obtained by the G.E. scientists on 21 December 1948. By dropping lumps of burning charcoal impregnated with silver iodide from an aircraft, approximately 6 square miles of supercooled stratus, having a thickness of 1000 ft and a temperature of $-10\,°C$, were converted into ice crystals by less than 1 oz. of silver iodide.

The first large-scale seeding trials using ground-based silver iodide generators are believed to have been carried out in Arizona and in the eastern part of Washington State in June 1950.

The experimental seeding of cumulus clouds

The most satisfactory method of seeding individual cumulus clouds, and of observing subsequent developments, is to disperse the seeding agent from an aircraft equipped with radar and under the control of a ground radar station which can keep it and the clouds under observation. In this way one can be quite certain of the artificial nuclei entering the cloud, and can observe changes in cloud structure, the onset of precipitation on the radar and when this falls from cloud base, all in relation to the time of seeding. Comparisons can also be made with neighbouring unseeded clouds.

The seeding of large supercooled cumulus clouds with dry ice was first reported in 1947 by Kraus and Squires working in Australia. They observed that six out of eight seeded clouds gave radar echoes indicating the presence of rain, and that in four of these, heavy rain reached the ground. The most spectacular case is illustrated in pl. XXI. A cloud with base at 11000 ft and summit at 23000 ft was seeded with 150 lb of dry ice, whereupon it grew to 29000 ft in 13 min. A radar echo appeared within 5 min of seeding, and after 21 min heavy rain was observed to fall from the base of the cloud. No other radar echoes were detected within a range of 100 miles.

Similar occurrences of rapid growth after seeding have been reported but such spectacular development, believed to be a consequence of the latent heat liberated by the transformation of supercooled water into ice, is rather rare.

Experiments in which a few pounds of dry ice crushed into pellets of about $\frac{1}{2}$ in. dia. are dropped into the tops of supercooled cumulus have now been carried out in many countries, for example, Australia, Canada, England, South Africa, and the United States. In general they agree in showing that there is a high probability of inducing precipitation with dry ice if the cloud summit is colder than -7 °C and the depth of the supercooled region exceeds about 4000 ft. From the results of more than 100 experiments performed in Australia during 1947–51 the following conclusions were drawn:

(*a*) With cloud-summit temperatures of -7 °C and colder there is a 100 % chance of producing precipitation. At temperatures between

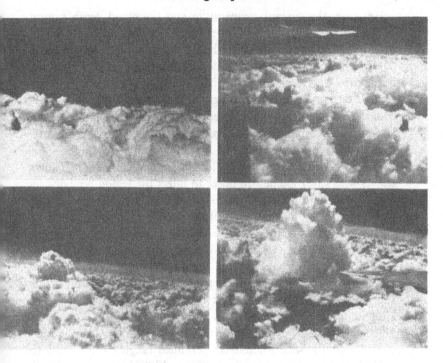

Plate xxi. The effect of seeding a cumulus cloud with dry ice (after Kraus and Squires). The top pictures show widespread cumulus with tops at 23 000 ft. Bottom left shows one cloud towering upwards 9 min after seeding. In bottom right, 13 min after seeding, it is seen reaching to 29 000 ft. Later it rose to 40 000 ft to form an anvil and rain fell for 2½ h. (Photograph by courtesy of Commonwealth Scientific and Industrial Research Organization, Sydney, Australia.)

−7 and 0 °C the chances of success fall off progressively, tending to zero at 0 °C. At temperatures of −15 °C and below the results lose their significance because the clouds have a high probability of raining naturally.

(*b*) The time at which precipitation appears at cloud base depends mainly on the thickness of the cloud; there is a gestation period of about 10 min, with an additional time of about 1 min for every 800 ft of cloud.

(*c*) The intensity of the precipitation from the cloud increases with increasing cloud thickness.

(*d*) A considerable fraction of the precipitation will reach the ground if the height of the cloud base is less than the cloud thickness.

However, the summit temperature and the cloud depth cannot be the only factors which will determine the chances of successful seeding: for example, the strength and duration of the updraughts, controlled by the convective processes, must be sufficient for the particles to grow to precipitation size. Herein lies a probable explanation for the fact that a similar programme of dry-ice seeding of super-cooled cumulus clouds in the central United States produced no evidence of modifying the cloud behaviour; these clouds were short-lived compared with those in the Australian experiments.

There are fewer data on the effect of seeding individual cumulus clouds with silver iodide. In order to infect the region bounded by the -5 and $-15\,°C$ levels in a cloud of cross-sectional area 5 km^2 with nuclei in an average concentration of 1/l., one would require 10^{13} nuclei. The quantity of silver iodide required to produce this number of effective ice nuclei will depend on their method of production, but if introduced from an aircraft, may be only a few grammes. In order to ensure complete and rapid glaciation of the supercooled part of the cloud it may be necessary to use a few kilogrammes.

In an Australian experiment conducted in the mid 1950s, super-cooled cumulus with summit temperatures ranging from $-2·5$ to $-16\,°C$ and depths ranging from 4000 to 17000 ft were seeded with about 10 g (or 10^{14} nuclei effective at $-10\,°C$) of silver iodide from an aircraft flying through the upper levels of the cloud or just below cloud base. Of those clouds which had summit temperatures below $-5\,°C$, 72% precipitated within 20–25 min of seeding, 21% evaporated, and 7% showed no discernible effects. At first sight, these results seem very similar to those described above for seeding with dry ice but further trials, in which the clouds were selected for seeding on a randomized basis, failed to produce such clear-cut results.

In the last few years it has become recognized that, in order to obtain a meaningful statistical evaluation of cloud-seeding experiments, it is necessary to eliminate, as far as possible, bias introduced by the operator in choosing suitable seeding occasions and by un-representative periods of weather. It has therefore become the practice to decide on the basis of the current meteorological situation whether, on a particular day, there will be cloud systems suitable for seeding, and then the actual decision to seed or not is made on the basis of an independent random choice determined in advance. For example, a number of 'Yes' and 'No' instructions, determined as the result of tossing a coin or by using a book of random numbers, are sealed in numbered envelopes before the experiment commences. On the first occasion designated as suitable for seeding, envelope no. 1 is opened and if the instruction is 'Yes', seeding is undertaken. If the instruction is 'No', no seeding takes place and the occasion is used as a control. In order to avoid the possibility of several 'Yes' or 'No' decisions in a row and to ensure that there are equal numbers of seeded and non-seeded occasions, one can make decisions only on alternate occasions. For example, whatever the randomized decision on the first occasion, one does the opposite on the second occasion, and a new decision is taken on the third, and so on.

A randomized experiment of this type, designed to evaluate the effects of silver-iodide seeding from aircraft upon individual cumulus clouds growing over Southern Florida, was carried out by Simpson and her collaborators in 1970. The sample contained 13 seeded clouds and 16 non-seeded or control clouds. All the seeded clouds achieved cumulonimbus status as did 10 of the control clouds. On six days the seeded clouds showed more vertical growth than did the control clouds, the differences on individual days ranging from 19000 ft to 5000 ft with an average of 10500 ft. On three days, however, the non-seeded clouds grew more than the seeded clouds by amounts ranging from 6000 ft to 15000 ft with an average of 4000 ft.

The rainfall from all clouds was estimated from the intensity of the echoes produced on a 10 cm radar. The authors claim that, during the 40 min following seeding, the total rainfall from the seeded clouds was, on average, almost double that from the non-seeded clouds, the excess amounting, on average, to about 100 acre-feet. It is difficult to assess the significance of this last figure because the area covered by the rainfall is not given, and the errors of the rainfall measurement which must have been quite large, probably at least $\pm 30\%$, are not estimated. Moreover the authors state that the seeded clouds produced more rain because they were bigger and longer-lasting than the control clouds and not because their rainfall intensity was significantly higher. Also, the size distribution of the raindrops from the seeded clouds was not noticeably different from that of the unseeded clouds. This raises doubts as to whether the seeding had any marked influence and whether the increased vertical growth and associated increase in rainfall may not have been a consequence of the seeded clouds being larger also in horizontal extent and therefore less attenuated by mixing with the surrounding air. Clearly the results of this experiment require much more careful examination and appraisal before they can be accepted at their face value.

Attempts to release showers by spraying cumulus clouds with water and aqueous solutions were made before the theory of the coalescence mechanism was clearly formulated. Experiments in which anything between 1 and 100 gallons of water in the form of fairly large drops are introduced into the tops of cumulus clouds have produced no marked effects. This is to be expected; that the spraying of a few gallons of water into the *top* of a cloud cannot produce a shower unless a process of raindrop multiplication (see page 109) occurs, is shown by the following simple calculation. If a shower cloud of base area 10 km² produces 5 mm of rain, the equivalent volume of water is about 10 million gallons. If this were released by the introduction of 10 gallons into the top of the cloud, the drops in this 10 gallons would have to grow one million-fold. But even if drops of initial diameter 0·5 mm were able to grow to the maximum stable size of 5 mm, this would represent only a 1000-fold increase in mass. It appears that the required growth could result only if the fragments of breaking drops grow, in turn, to break-up size and cause a rapid multiplication of raindrops. However, clouds deep enough to support such a chain reaction will usually have a high probability of precipitating naturally.

A much more efficient method is to introduce small droplets of radius 30–40μ into the *base* of a growing cloud and to capitalize on their subsequent growth during both their upward and downward journeys through the cloud.

In 1952 eleven experiments along these lines were undertaken in Australia in which water drops of median radius 25μ were sprayed at the rate of about 30 gallons per minute during flights made at about 1000 ft above cloud base. On six of the seven occasions, when the cloud thickness was less than 5000 ft, light rain fell out of the cloud but evaporated before reaching the ground. Neighbouring untreated clouds did not precipitate. In the four cases where the cloud depth exceeded 5000 ft (1·5 km) there was a considerable fall of rain or hail within a short time of seeding and on three of these occasions the neighbouring unseeded clouds did not rain. Calculations indicate that the introduction of 25μ-radius droplets into the base of a cloud having base temperature 10 °C and a steady upcurrent of 1-2 m/s could release a shower from a cloud of depth greater than 1·5 km—result which agrees well with these experiments.

Similar tests have been carried out in the Caribbean area in which water spray in quantities of about 450 gallons per mile was introduced into the tops of tropical cumulus clouds. Clouds were selected in pairs and one of each pair, chosen at random, was seeded. About one half of the treated clouds developed radar echoes, which were evidence of precipitation, while only about one quarter of the untreated clouds developed echoes. Similar experiments carried out in the mid-western United States were quite inconclusive. On the basis of these rather few data, it appears that the method of spraying *large* drops into the tops of convective clouds appears inefficient and too costly for large-scale application. Economically the introduction of *much smaller* droplets into the bases of the clouds seems more promising but the technique requires much more careful investigation.

In view of the important role which giant hygroscopic nuclei were believed to play in initiating the coalescence process, it appeared even more economical to use such nuclei rather than water droplets as a seeding agent. Since dry salt crystals of diameter 10μ will more than double their size while being carried up through the first few hundred metres of cloud, the dispersal of about 100 g of salt would be equivalent to that of about a gallon of water in 50μ dia. droplets. This method has been tried, mainly in East Africa. Bombs containing gunpowder and sodium chloride, carried aloft by balloons, were arranged to explode near cloud base and disperse salt particles of diameter $5-100\mu$. For the 38 days when salt was released, the total rainfall in an area 6-12 miles downward of the release point was up to 6 in. in excess of that for intermediate unseeded days, but the rainfall over an area extending 5 miles *upwind* was also greater by 2 or 3 in. It appears, however, that in these, as in other experiments carried out in Pakistan with salt dispersed from a ground generator, the number of salt particles reaching the clouds was insufficient to produce a detectable amount of rain even if each grew into a large raindrop. There is accordingly no convincing evidence that salt-seeding has produced positive results, and again a much more critical evaluation of the method is required.

The experimental seeding of layer clouds

In seeding supercooled stratiform clouds from aircraft, the usual procedure has been to drop granulated dry ice at the rate of a few pounds per mile of flight along a well defined track into the top of the cloud deck, and to photograph subsequent developments from above. The clouds are usually too thin to produce precipitation at the ground but seeding often produces spectacular changes in the appearance of the cloud. Along the seeded track the cloud becomes completely transformed into ice crystals. Gradually the turbulent motions in the cloud layer diffuse these crystals laterally so that after a few minutes a long lane a few hundred yards broad becomes affected. After a while, the crystals at the edge of the lane become so diffused that each has a large share of the available water and grows large enough to fall out of the cloud layer into the drier air beneath. The crystals then evaporate, and the cloud disappears in a broad strip around the seeded track to leave a long hole in the cloud.

Plate XXII shows the result of such an experiment carried out by the American Project Cirrus in November 1948. An extensive layer of stratus with base at 6000 ft, temperature -3.5 °C, and top at 7300 ft, temperature -5 °C, was seeded with dry ice at the rate of about 0.7 lb/mile. The seeded track took the form of a letter γ with a total length of about 57 miles. The photograph shows the appearance of the pattern 36 min after the loop in the upper part of the picture was seeded. At this time the width of the track to the right was 1.7 miles. An observer on the ground saw curtains of snow falling from the edge of the open lanes. Ultimately new clouds developed in the openings, so that within 2 h they were more than half filled, but even after 4 h some openings were still visible.

Many similar experiments have been carried out in Canada, England and elsewhere with similar results. Whether or not a hole develops depends upon the rate of seeding, the temperature, depth and water content of the cloud, the extent to which air is converging or diverging into the region, and the intensity of lateral mixing, the area affected depending largely on the last two factors. In a favourable situation the seeded lanes may spread to widths of 2 miles in half an hour.

A rather extensive series of trials conducted in the Maine region of the United States has shown that cloud decks less than 3000 ft (1 km) thick and colder than -5 °C could be cleared equally well

Plate XXII. A γ-pattern of ice crystals produced in a thin, supercooled cloud layer by seeding with dry ice. The ice crystals transported to the edges of the seeded lane grew large enough to fall out of the cloud, leaving a long rectangular hole which was nearly 2 miles wide when the picture was taken 36 min after seeding. (Photograph by courtesy of General Electric Company, Schenectady, U.S.A.)

with seeding rates between 1 and 10 lb of dry ice per mile, but thick, convective layers were sometimes under-seeded with rates as high as 10 lb/mile. When clouds only 500 ft in depth were seeded only about 10 min elapsed before the ground could be seen, but for depths of 2000 ft this time was increased to about 25 min.

Similar results have been obtained with silver iodide and suggest a technique for dispersing layers of supercooled fog in cold weather. Fogs at temperatures below $-5\,°C$, when silver iodide may be effective, are very rare in England, but the possibilities of dispersing supercooled fogs by mobile silver-iodide generators should be more actively explored in places where these are relatively frequent.

Large-scale cloud-seeding operations

The experiments just described have clearly demonstrated that the introduction of seeding agents into suitable clouds is often followed by observable, and sometimes spectacular effects. They do not, how-ever, tell us very much about the possibilities of modifying the rainfall from widespread cloud systems extending over thousands of square miles. For operations on this scale, it is the general practice to release silver iodide in the form of a smoke from ground generators producing about 10^{13} nuclei/s, relying on the air currents to carry it up into the supercooled regions of the cloud. Trials of this type have been carried out on a large scale during the past 20 years, mainly in the United States, Canada and South America by commercial operators, employed by ranchers, farmers, power companies, and public utilities.

The generators are switched on when the arrival of suitable storms in the target area is imminent and kept burning during their passage, the working hypothesis being that seeding may cause an increase in the precipitation over that which would fall naturally. In evaluating the results of such trials, the vital question is: is it probable that the precipitation pattern which appeared immediately following seeding would have appeared even if no seeding had taken place? The difficulty lies in the inherent variability of natural precipitation and the meteorologists' present inability to predict with sufficient accuracy what would have occurred in the absence of treat-ment. In places having a fairly plentiful rainfall of 40 in./year, the rainfall in any one year may easily deviate from the average over a 50-year period by 10 or 15 %, while in dry regions like Arizona, the variation may be as much as 50 %. It is in such dry regions, where rainmaking is most in demand, that it is most difficult to assess the efficiency of cloud seeding because of the large year-to-year variation in the natural rainfall.

Since we are still far from being able to form an accurate estimate, on a purely physical basis, of how much rain will fall naturally in a given region, the effects of seeding must necessarily be assessed statistically. The target area having been decided, the usual procedure is to choose an adjacent control area, preferably of much the same size, shape and topography and unlikely to be affected by the seeding agent. One method of evaluation is then to plot, for past years, the annual target-area rainfall against that for the control area and, knowing the rainfall in the control area during the

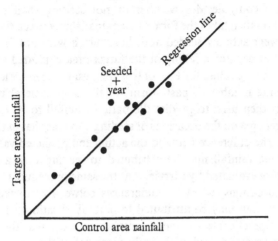

Fig. 28. A typical correlation between the rainfall in the target and control areas of a cloud-seeding operation.

period of seeding operations, use this graph to predict the natural rainfall in the target area. The difference between the latter and the actual amount is then attributed to the effects of seeding with a degree of confidence determined by the scatter of the historical data on the graph (see fig. 28). In other words, the assessment takes the form of determining whether there are significant departures from the normal historical relationship between the rainfall in target and control areas and, clearly, an apparent departure can be statistically significant only if there is a high correlation between past rainfall in the two areas.

Although this method of evaluation may seem very reasonable it is open to the serious criticism that if the pattern of major storms during the seeding period differs appreciably from the long-term average pattern on which the historical relationships are based, then the latter cannot be used to predict accurately the rainfall in the target area, and consequently the assessment of seeding effects may be most misleading. For these reasons statisticians have insisted that an acceptable form of randomization should be incorporated into the operation from the beginning; for example, seeding should be carried out on only half the suitable occasions selected at random, the other half being used as a control. After it has been decided on meteorological grounds that a storm is suitable for seeding, the result of a previously performed randomized experiment (such as tossing a coin) decides whether or not seeding shall take place. Evaluation then takes the form of a comparison between those storms which were actually seeded and those which were not. The rainfall per day or per storm falling in the target area is plotted against the corresponding values for the control area, using the historical records for a large number of past storms of the same general type. This graph is then used to predict the natural rainfall to be expected in the target area on the occasions of seeding knowing that in the control region. The difference between the actual and predicted values of the target-area rainfall may be attributed to seeding with a degree of confidence evaluated by carrying out the same procedure for the non-seeded occasions, when the differences between predicted and observed rainfall may be attributed to natural variations. If it should prove impracticable to find a suitable control area, the unseeded occasions may be used as the only control, but then the experiment will have to be carried out for a longer period in order to achieve the same degree of significance in the results.

In another variant, the so-called cross-over design, there are two target areas, say A and B, sufficiently close to ensure a high correlation between normal precipitation amounts, and yet sufficiently distant so that seeding in A is unlikely to influence the precipitation in B and *vice versa*. Seeding is done on every suitable occasion, but the decision as to which of the two targets A and B is to be seeded is determined by a randomized procedure. This cross-over design is the most efficient of the procedures just described, provided that seeding in one target area does not influence the precipitation in the other. On the other hand, if there were strong interaction between the two areas, the conclusions drawn from the analysis of the cross-over design could be misleading or erroneous.

Unfortunately randomized procedures have not been adopted in the great majority of commercial rain-making operations and so it is not possible to assess their results on an objective scientific basis. Accordingly we shall not attempt to summarize the vast literature in this field, but rather select for description and discussion a representative group of carefully-conducted randomized large-scale operations that will not only illustrate the operational and evaluation procedures, but also give a fairly balanced picture of the present state of the art.

Seeding with silver iodide from aircraft

(i) A group of operations conducted in Australia

A 5-year randomized operation was carried out from June 1955 to the end of 1959 in the Snowy Mountains of Australia. The objective was to determine whether seeding with silver iodide can produce an economically significant increase in precipitation over the Snowy Mountains at about 6000 ft altitude where it can be utilized for generating hydroelectric power. Aircraft seeding was employed to make sure that the silver iodide actually entered the cloud systems involved, the output of the airborne generator being about 10^{17} nuclei/h, active at -17 °C. Cumulus clouds were seeded at near the -6 °C level. The central target area was only 35 square miles but a surrounding area of 1100 square miles was thought to be affected by the seeding. The control area was 750 square miles. Seeding was confined to the non-summer months when the dominant cloud systems were associated with cyclonic and frontal disturbances, the vertical motion of the air being intensified by orographic lifting. Operations were divided into 'seeded' and 'unseeded' periods each of not less than 8 days' duration, the division being made on the basis of a set of random numbers. This series was unknown to the individual who decided the end of each period and the beginning of the next on the basis of the passage of successive anticyclones across the region. Analysis of the results of the 5 years' operations was based on the total precipitation of the three areas as estimated by integration of isohyetal maps prepared from rain-gauge data. Over the 5 years, the ratio of the total precipitation in the target area to that in the control area during the seeded periods exceeded by 19% that in the corresponding unseeded periods. The excess was judged significant at the 3% level according to one statistical test, and at the 9% level according to another. But, for individual years, the excess of target over control-area rainfall varied from 3 to 27%. The corresponding figures for the 'seeded' area surrounding the target area were -4 to $+19\%$, with a 5-year average of 11% judged to be significant at either 4 or 24%, according to the nature of the test. In view of these figures, the strong differences between the rainfall régimes in the different years, the fact that the target area is at a higher elevation and has a considerably higher natural spring rainfall than the control area, and because of the difficulty involved in measuring the precipitation, much of which was snow driven by strong winds, the overall result of the operation must be judged inconclusive.

In three other locations, South Australia, New England, and the Warragamba Catchment, the target-control cross-over design was used, the random process determining which of two areas should be seeded, and which should not, during each period. The effect of seeding was calculated from the ratio

$$\frac{(ps)}{p(ns)} = \left\{ \frac{p(A, s)}{p(A, ns)} \times \frac{p(B, s)}{p(B, ns)} \right\}^{\frac{1}{2}}, \qquad (6\cdot1)$$

where $p(A, s)$, $p(B, s)$ are respectively the average totals of precipitation falling in areas A and B on days of seeding, and $p(A, ns)$, $p(B, ns)$ the corresponding totals on unseeded days. Analysis of the results indicated that seeding produced a rainfall increase of only 4% in New England and small decreases in the other two experiments that were hardly significant and may well have occurred by chance.

Pointing out that the largest apparent increases in precipitation due to seeding tend to occur in the early years of an operation, and are often offset by apparent decreases in latter years, Bowen has suggested that seeding may produce persistent effects that may carry over from one year to another.

(ii) Seeding of orographic cumulus in Arizona

Two experiments, conducted to investigate the possibility of modifying the precipitation from cumulus clouds that form almost daily in summer over the mountains near Tucson, Arizona, have been described by Battan and his colleagues. The seeding was randomized by pairs of days, i.e. when a forecast based on objective criteria indicated that the following two days would have clouds suitable for seeding, the decision whether the first or second day should be seeded was made by a random process. In the first experiment, in 1957–60, silver iodide smoke was released from an aeroplane flown at the -6 °C level along a line perpendicular to the wind, upwind of the target area. Rainfall was measured by a network of 29 recording rain-gauges during a 5-hour period that included the seeding period lasting 2–4 hours. Increases in rainfall were indicated in the first two years, and even larger decreases during the last two years. All the data for the whole 4-year period indicated an overall decrease of 30%, but at a level of significance that suggested that this result could have occurred by chance. Indeed, much of the apparent large decrease for 1959 was due to extremely large falls of rain on one non-seeded day; if this pair of days is excluded from the analysis, the indicated overall decrease over 4 years falls to only 7%.

Following a preliminary analysis of the first experiment, a new experiment was started, using more restricted criteria for selecting the pairs of days in order to reduce the frequency of experimental days with zero rainfall, seeding now being carried out at a few hundred metres below cloud base, instead of at the −6 °C level. The target area was reduced and the number of rain-gauges in it was increased. The results of this experiment, which was carried out in 1961, 1962 and 1964, were much the same as those of the first, and indicate a net decrease of 30% in rainfall over the 3-year period but, with significance probabilities estimated to lie between 0·16 and 0·30; this could well have occurred by chance.

(iii) Project 'Whitetop' (1960–4), Missouri, U.S.A.

This was a carefully designed long-term experiment to test the effect of seeding with silver iodide from aircraft on the release of precipitation from non-orographic summer cumulus clouds in Missouri. The days were identified as seedable on the basis of objective criteria indicating the likelihood of instability showers. If the quantity of precipitable water in the atmosphere shown by the morning ascent of three nearby radio-sonde stations exceeded 3 cm, and the wind direction at 1300 m was between 170° and 340°, the day was designated operational, and a random process, determined in advance, decided whether seeding should take place or not. On seeding days, aircraft equipped with silver iodide burners flew to and fro along a seeding line 30 miles long and 45 miles upwind of a central radar site, the smoke being released at the level of cloud base from noon until 1800 Local Time. On both seeded and unseeded days, this line was used to define 'Chicago-plume' positions on the basis of the most divergent winds between the seeding level and 4500 m. Narrower plumes, based on the winds at the seeding level, were called 'Missouri plumes'. The target area was considered to be the area covered by these plumes, and comparisons were made between the precipitation in the plumes on seeded and on unseeded days, and between precipitation in the plumes on seeded and on unseeded days, and between precipitation in the plumes and in the remainder of the circular experimental area, 60 miles in radius, which was labelled 'out of plume' area and considered to be uncontaminated by the silver iodide.

The rainfall data were obtained from a rather sparse network of rain-gauges (one gauge per 270 square miles) and subjected to three independent statistical analyses. All the analyses, though treating the data rather differently and applying different tests of significance, nevertheless agree that there was a significant *decrease* in the precipitation on seeded, as opposed to unseeded, days both in the Missouri plume and the Chicago plume, amounting to 60% and 40% respectively. Precipitation was also less in the 'out of plume' area of $> 10^5$ square miles on the seeded days, but this result carries a lower level of statistical significance. A later study, in which the rainfall analysis was extended up to a radius of 180 miles, found an overall decrease of about 20% on seeded days averaged over the entire area, but the significance probability was only 0·13.

Finally, the most recently published statistical study states 'any conclusions about the effectiveness of seeding, one way or the other, that are based on the Whitetop experiment must be made with extreme caution'.

(iv) The Israeli experiment

A randomized cross-over experiment, involving the seeding of clouds, mostly cumulus, with silver iodide from aircraft, was carried out in Israel during 1961–6. The trial involved two experimental areas, each about 50 km in maximum dimension, for which the correlation between daily amounts of precipitation was 0·81. They were separated by a buffer zone about 20 km wide. Silver iodide was released at the rate of 800 g/h from a single aircraft flying just below cloud base in an area displaced upwind of the target by a distance equivalent to 0·5 h run of the wind. Each day, one area was designated for seeding on the basis of a randomized decision taken in advance, but only if suitable clouds were present was seeding actually carried out.

The effect of seeding was expressed by the ratio of the average daily precipitation falling on seeded and non-seeded days and was calculated from (6·1). The seasonal values of this ratio, calculated from data on all the experimental days, varied from 1·01 to 1·65. If only rainy days, defined as days on which rain fell in the buffer zone, were included in the analysis, the seasonal ratios ranged from 1·02 to 1·74. Over the 5½-year period as a whole, the statistical analysis indicated a rainfall increase of 18%. Most of this increase could be attributed to large effects occurring on only a few days. This is one of the very few long-term randomized experiments that appear to have produced significant increases in rainfall.

The Israeli scientists carried out a number of statistical tests on their data but found no evidence for the persistence of seeding effects, either from day to day, or within each season, or from season to season.

Seeding with silver iodide from ground generators— Project 'Grossversuch III'

One of the very few well-designed large-scale operations using ground-based silver iodide generators was the Swiss project 'Grossversuch III', whose primary object was to reduce the incidence of large damaging hail from cumulonimbus clouds growing over the southern slopes of the Alps.

In the experiment, which was fully randomized, experimental days were those during the summer months of the seven years 1957–63 for which thunderstorms were forecast on the previous afternoon. The decision whether or not to seed on a particular experimental day was taken by selecting at random an envelope containing a 'Yes' or 'No' instruction. In this way, 292 experimental days were selected during the seven years, of which 145 were seeded. The seeding was carried out by 20 silver iodide generators operating from 0730 to 2130 hours on a pulsating schedule of 5 min on and 10 min off. The generators were located in, and to the south of, the 3500 km target area ranging in altitude from 200 to 3400 m.

The results showed no significant difference in the duration, areal extent, or intensity of hail on seeded and unseeded days, but the frequency of hail on seeded days was considerably greater than on unseeded days, with a probability of only 4% that this difference could have occurred by chance. As far as precipitation was concerned, however, the average rainfall on seeded days was 21% higher than on unseeded days, with a significance level bordering on the acceptable. The evidence also seems to suggest that seeding may have produced quite large increases in rainfall at distances of up to 100 miles from the target area. This is by no means proved, but should seeding be capable of producing such long-range effects as suggested both here, and by the analysis of Project Whitetop, the results of experiments employing the cross-over design would be open to serious doubt.

Suppression of large hail

In recent years, projects aimed at suppressing large, damaging hailstones have been carried out in several countries, for example Austria, Argentina, Bulgaria, Canada, France, Italy, Yugoslavia, Kenya, Switzerland, Russia and the United States. The best designed experiment, in that it was properly randomized, was the 'Grossversuch III' project in Switzerland described above. The most intensive operation was mounted in Kenya, where 10000 rockets, each containing about 800 g of TNT, were exploded to treat a total of about 150 storms in an area of only about 1500 hectares. The rockets contained no nucleating agent, but it has been suggested that the shock waves from the explosion might shatter the large hailstones into smaller, less damaging fragments, or produce large numbers of ice crystals by adiabatic expansion and cooling of the air. None of these three projects nor some others for which detailed reports are available provide convincing evidence of success in reducing either the frequency or intensity of hail. Indeed the scientists in charge of the operations in Switzerland and Argentina considered them failures, while the National Science Foundation annual reports on weather modification have so far made no claim for success in the United States.

The working principle behind most of the trials is that, if a growing supercooled cloud is seeded with a massive dose of ice nuclei, the competition for the available water between the high concentration of ice particles will prevent any of them growing into large hailstones. One of the major doubts about most of the practical trials, especially those employing ground generators, is whether artificial nuclei were able to reach the supercooled parts of the cloud in anything like sufficient concentrations. The most impressive claims for success come from the Soviet Union, where large-scale operations have been conducted in Georgia, Armenia, and the Caucasus since 1962, and in Moldavia since 1964.

Soviet scientists believe that an important feature of a hail-producing cloud is the formation of an 'accumulation zone' where the concentration of liquid water becomes very high as the result of the updraught reaching its maximum speed at this level. If the accumulation zone reaches above the 0 °C isotherm, hailstones are assumed to originate by the freezing of supercooled raindrops and thereafter are able to grow to diameters of 3 cm within 4–10 min, depending on the concentration of supercooled water and the concentration of particles competing for it. There is, in fact, no real evidence for the existence of persistent accumulation zones and it is not clear how they could be maintained for several minutes once precipitation has developed.

Nevertheless, the technique of delivering the seeding agent into the heart of an incipient storm is sound, and the organization and logistics of the Soviet experiments are impressive. They aim to introduce silver iodide in the clouds between the -6 and $-12\,^\circ\mathrm{C}$ levels, either by rocket, as in the Georgian operations, or by artillery shell, as in the Caucasus. The shells, 12·5 cm in diameter and weighing 33 kg, have a maximum range of 15 km and contain 100 g of silver iodide which yields about 10^{15} ice nuclei active at $-10\,^\circ\mathrm{C}$. Field tests show the nuclei to be dispersed through a volume of order 10 km^3 to give an average concentration of order 100/litre. One shell is fired for every 5 km^3 of the estimated volume of the supercooled part of the cloud, so that a large cloud may receive 1 kg of silver iodide within about 1 h. Operations in the Caucasus employed 16 artillery batteries in 1969 to protect an area of 5 million hectares. The relative advantages of shells and rockets are a matter of some dispute, but the rockets have only half the range of the shells and cannot be controlled with the same degree of accuracy.

The probability that large hail will develop is assessed on a number of criteria based on radar measurements. These include the height and temperature of the top of the radar echo, its vertical extent, and maximum reflectivity, and the relative vertical depths of the echo above and below the 0 °C level but it is not known how these various factors are weighted. It is claimed that by measuring the reflectivity of the radar echo on two wavelengths, 3·2 cm and 10 cm, simultaneously, it is possible to detect the presence of large concentrations of hail greater than 0·6 cm in diameter and to estimate their average diameter on the assumption that they are mono-disperse, spherical, and wet. In practice, none of these three assumptions is likely to be generally true and consequently it is difficult to obtain reliable information on hailstone sizes from radar data alone.

None of the Soviet experiments is randomized, so the results are assessed on subjective judgements of crop losses occurring in the protected regions compared with those in unprotected control regions. It is claimed that losses have been reduced by as much as 80–90 % in the Caucasus and Georgia, with economic benefits estimated at between 10 and 40 times the cost of the operations. These claims are difficult to assess in the absence of proper statistical tests, and in view of the fact that suitable comparison areas are very difficult to obtain.

Discussion

Looking then at the world-wide attempts to achieve some measure of weather modification and control, one can find little convincing evidence that large increases in rainfall can be produced consistently over large areas. Although it is possible that marginal but still economically important effects of, say, 10–20% may be produced, in general it has not been possible to distinguish induced changes of this magnitude from the natural variations in rainfall.

Apart from doubts about the validity of the evaluation procedures that have been used to assess massive cloud-seeding operations, these are conducted on certain basic assumptions that are insecurely based, as may be illustrated by our present inability to answer the following important questions.

(*a*) How often in a given locality, and under what conditions, are natural ice nuclei deficient, and when and in what quantities should they be supplied artificially?

(*b*) In which part of the cloud or cloud system, and at which stage of its development, should the seeding agent be introduced in order to stimulate the maximum effect? May not seeding sometimes stimulate the premature release of precipitation, suppress the natural growth of the cloud, and thereby reduce rather than increase the precipitation from it?

(*c*) When silver iodide smoke is released from the ground, in what concentration does it reach the supercooled regions of the cloud and how will this vary with the meteorological situation and the distance from the source?

(*d*) For how long does the nucleating agent remain effective under specified atmospheric conditions?

Even more important, it is difficult to escape the conclusion that, once initiated, the distribution, intensity, and duration of precipitation are largely controlled by the cloud dynamics, and influenced by air motions on scales both larger and smaller than that of the cloud itself. Perhaps the greatest weakness in our present knowledge and approach to cloud modification is our poor appreciation of the relative importance of microphysical and dynamical processes in controlling the efficiency of natural precipitation mechanisms.

At the present time, when the efficacy of large-scale seeding has yet to be firmly established despite 25 years of effort, one can only indicate some of the possibilities that may arise from improved knowledge and techniques.

In middle latitudes, much of the precipitation falls from deep layer-cloud systems whose tops usually reach to levels at which there are abundant natural ice nuclei and in which the natural precipitation processes have plenty of time to operate.

Here it is relevant to mention the outcome of Project Scud, in which massive seeding with both dry ice and silver iodide was conducted over the east coast of the United States during the winters of 1953 and 1954 to determine the effects of seeding on the development of cyclones. On 19 out of 37 occasions of potential cyclo-genesis, seeding was carried out on a randomized schedule, the other 18 occasions being used as controls. Altogether, 250 lb of silver iodide and 30 tons of dry ice were used in the operation. A careful statistical analysis revealed that if seeding produced any effects on either the total precipitation over the area or on the sea-level pressure field, they were too small to be detected against the background of the natural variance. A similar result also emerged from an operation in the State of Washington where the seeding, with dry ice, of migratory cloud systems associated with cyclonic activity produced no detectable effects. Indeed favourable seeding opportunities were rather infrequent because of the abundance of natural ice crystals in the tops of the deep cloud layers.

The prolonged, steady release of precipitation from the deep clouds of a well-established warm front, in which there is little or no storage of supercooled water above the -5 °C level, suggests a quasi-steady state, dominated by the dynamics, in which the processes of nucleation and particle growth adjust themselves to match the rate at which moisture is released by the vertical motion. If this be so, seeding is unlikely to have a major effect on the intensity or duration of the precipitation from these mature storms. It is possible, however, that judicious seeding of the cloud during the early stages of its development, when thick layers of supercooled or mixed cloud may have already formed, could forestall the natural release of precipitation and thereby effect some redistribution on the ground. But the scope seems rather limited and an elaborate surveillance system would be required to achieve the best timing of such an operation.

Perhaps more promising as sources of additional rain or snow are the persistent, supercooled orographic clouds produced by the ascent of damp air over large mountain barriers. The continuous generation of an appropriate concentration of ice crystals near the windward edge might well produce persistent light snowfall to the leeward, since water vapour is continually being made available for crystal growth by the lifting process. The condensed water, once converted into snow crystals, has a much greater opportunity of reaching the mountain surface without evaporating, and might accumulate in appreciable amount if seeding were maintained for many hours.

Rather different considerations apply in the case of incipient shower clouds where seeding could produce quite opposite effects depending upon the dynamical structure of the cloud, its state of organization and development, and such important environmental factors as the wind shear and humidity. In a dry environment, potentially unstable over only a limited depth, the cloud may have a high probability of dissipating without precipitating naturally; in such a case, acceleration of particle growth by seeding may increase the probability of releasing a shower, albeit only a light one. On the other hand, if a cloud growing in a moist, highly unstable environment is seeded early in its career, the premature release of precipitation may well destroy the updraught and cause the cloud to dissipate before reaching its maximum potential. This latter possibility has not received sufficient attention, but it may account, at least in part, for the fact that several large projects, for example, in Australia, Arizona and Missouri, involving the seeding of cumuliform clouds, have led to an apparent *decrease* in the rainfall over the target area.

Another possibility is that glaciation following seeding may lead to enhanced growth of the cloud stimulated by the release of latent heat. Although explosive growth of the cloud, in the manner observed in the famous early experiment of Kraus and Squires in 1947, is a rather rare event, observations such as those of Simpson mentioned on page 129 suggest that modest growth may quite often follow seedling.

The feasibility of decreasing rainfall by 'over-seeding' clouds has also to be considered. If the concentration of ice crystals in a cloud were to exceed $1/cm^3$, none would be able to grow sufficiently large to fall out of the cloud and reach the ground as precipitation. The introduction of an excessive concentration of ice nuclei into supercooled clouds might therefore retard or prevent the development of precipitation and suppress the formation of large hail and lightning.

It is unlikely that 'overseeding' has been regularly achieved in any large-scale operation using ground-based generators; indeed it seems impracticable to ensure persistent over-seeding of active, moving storms by this means. Delivery of the seeding agent directly into the core of the cloud by rocket, shell, or aeroplane seems the only feasible method.

The greatly exaggerated claims made by the early cloud-seeding operators are now largely discredited; current claims are much more modest and the difficulties of executing and evaluating cloud-seeding operations are now much more clearly and widely recognized. Statistical assessments of these operations are continually revealing features in the rainfall patterns that are difficult to interpret and explain, even qualitatively, in meteorologically convincing terms. The formulation of reliable numerical models of cloud growth and development, capable of predicting important and readily observed characteristics of seeded and unseeded (control) clouds, offers the best long-term prospect of assessing the potentialities and results of cloud-modification experiments, and should provide deeper physical insight into the problem than the purely statistical approach. However, it would be optimistic to expect the early development of models capable of predicting precipitation amounts with sufficient accuracy to detect the changes of 10–20% that are currently being attributed to cloud-seeding operations. Assessment of such experiments will therefore continue to depend on a combination of statistical and physical criteria, but it is on the latter that the greatest effort is now required. Little further progress is likely until we have acquired a much deeper understanding of the physical, and particularly the dynamical processes that control the release, duration, and intensity of precipitation from the major types of cloud systems.

THE ELECTRIFICATION OF THUNDERCLOUDS

Introduction

Ever since Benjamin Franklin and the French physicist d'Alibard proved, in 1752, the presence of electricity in thunderclouds, its origin and manifestation in the lightning flash have provided challenging scientific problems which still await solution.

Franklin's researches had been mainly concerned with the manner in which sharp points of metal could remove or discharge electrified bodies. Also, using a frictional electrical machine of his own design, he had made sparks a few inches long and commented upon their similarity to the lightning discharge. He went on to add: 'The electrical fluid is attracted by points. We do not know whether this property is in lightning. But since they agree in all particulars wherein we can already compare them, is it not probable that they agree likewise in this? *Let the experiment be made.*'

So Franklin was the first to propose a direct test of whether 'clouds that contain lightning are electrified or not'.

In a letter to the Royal Society in 1750 he proposed that a long, pointed iron rod, supported by an insulating glass stool, be mounted in a small hut on a high chimney or tower. He suggested that, if the insulator be kept clean and dry, a man standing on the stool during the passage of a thunderstorm might become electrified because the rod would attract electricity to him from the cloud. Alternatively he suggested that the man might stand on the floor of the hut and bring an earthed metal loop attached to an insulating handle near the rod and then electric sparks might be expected to jump from the rod to the wire.

At this time there was no very high steeple in Philadelphia with which Franklin could test his proposals, but his ideas were eagerly pursued in France where the physicist d'Alibard decided to try the experiment without a tower. He used a 40 ft iron rod tied to wooden masts by silk ribbon and a glass bottle as an insulator. The equipment was placed in charge of an old soldier named Coiffier who, on 10 May 1752, at Marly, near Paris, brought an earthed wire near to the rod while a thunderstorm was overhead and obtained the expected stream of sparks and so the first direct proof that thunderclouds were electrified.

Meanwhile, Franklin, without knowing of the success of the French experiments, had decided to try it with a kite made from a silk handkerchief attached to a length of string ending in an insulating silk ribbon. When a thunderstorm approached he observed some of the fibres on the string to stand erect; discerning this as a sign of electrification he put his knuckle to a key attached to the string and produced a considerable spark. This happened in June 1752, a month after d'Alibard's experiment.

Franklin next decided to determine the sign of the charge on thunderclouds. On 12 April 1753 he successfully collected electricity from an insulated rod and stored it in a Leyden jar and then compared it with charge of a known (positive) sign from a rubbed glass rod. On the basis of this and later experiments he concluded that '*the clouds of a thundergust are most commonly in a negative state of electricity, but sometimes in a positive state—the latter, I believe, is rare*'.

This statement of Franklin's remained the only direct and reliable information on the subject for 170 years and even today we would wish to modify it only to the extent of replacing 'clouds' by 'bases of clouds'.

For a more detailed account of these pioneer experiments and of the fascinating history of the lightning conductor the reader is recommended to read *The Flight of the Thunderbolts* by B. F. J. Schonland (Oxford University Press, 1950).

The thunderstorm as an electric generator

In essence, we can regard the thundercloud as an electrostatic generator which produces electrical charges, both positive and negative, and separates these so that positive charge becomes concentrated in one region of the cloud and negative charge in another region. As separation of the positive and negative charges proceeds, the electrical field between these oppositely charged regions, or between one of them and the earth, grows until electrical breakdown of the air occurs. A conducting channel is then produced along which at least part of the separated charge disappears during the passage of a lightning flash. Further separation of charge now takes place and the same sequence of events is repeated.

In principle, then, the operation of this 'natural dynamo' is simple, but the detailed mechanisms of charge generation, separation and dissipation are complex and still not completely understood. Before discussing various theories of these processes, we shall review the main facts about the thunderstorm and its electrification as revealed by observation and measurement, and from them deduce the conditions which any satisfactory theory must obey.

The meteorological structure of a thunderstorm

Exploration by means of aircraft and radar reveals that a thunderstorm consists of one or more active centres, or *cells*, which contain strong vertical air currents and are the locale of raindrop and hailstone formation and of lightning activity. Each cell goes through a fairly well defined life cycle, comprising a growth stage, a mature stage and a decaying stage. While the whole life of a cell may occupy about an hour, the mature stage, during which precipitation is released and the lightning occurs, lasts for about 15 to 20 min. Very large, persistent thunderstorms may develop from successive cells developing in turn.

A mature cell (or that part of it in which precipitation particles are detected by radar) may have horizontal dimensions of 1–5 miles and extend vertically from the cloud base to a level at which the temperature is $-40\,°C$, a depth of typically 25000 ft (7·5 km); the visual cloud will usually be surmounted by a vast anvil-shaped mass of ice crystals reaching to even greater heights. The vertical air currents in a mature thunderstorm may exceed speeds of 30 m/s (60 m.p.h.), although values of 5–10 m/s are more common.

The bases of thunderstorms are usually warmer than $0\,°C$, so that in the early stages of development, water drops will predominate in the lower regions, while a mixture of ice crystals and supercooled droplets will appear at colder levels up to the $-40\,°C$ level, above which only ice crystals exist. The growth of raindrops by coalescence between water drops of different sizes and falling speeds commences in the lower part of the cloud, while the capture of supercooled droplets by ice crystals initiates the formation of pellets of soft hail at higher levels. As we have seen on page 116, some of these ice pellets may later develop into true hailstones but, although intense storms giving large hail are almost invariably accompanied by lightning, the presence of *large* hail does not appear to be a frequent or necessary condition for the onset of lightning. The growth of the precipitation particles continues until their size and concentration become so large that they can no longer be supported by the upward air currents and they begin to fall towards the earth. The downward rush of heavy precipitation helps to destroy the upcurrent, a down-current now being initiated instead; this marks the onset of the mature stage of the storm with which we shall now be mainly concerned.

The electrical structure of a thunderstorm

In fine, undisturbed weather, the atmosphere exhibits a fairly uniform, steady, downwardly directed electrical field due to the existence of a negative charge on the earth's surface and a net positive space charge in the atmosphere. The intensity of the vertical electrical field has a maximum value at the ground where its magnitude is 120 volts per metre (V/m) when averaged over the whole earth, and 130 V/m over the oceans; in industrial regions where the air is highly polluted, the field may be considerably enhanced, the mean value at Kew Observatory being about 360 V/m. The field intensity (or potential gradient) decreases at greater heights, at 10 km falling to only 3 % of its surface value. The potential of the atmosphere with respect to the ground increases with altitude up to about 20 km, above which it remains nearly constant at about 400000 V. The very small potential gradients which exist above 20 km indicate that the air at these levels is highly conducting.

This so-called 'fine-weather' electric field is violently disturbed in the presence of storms and particularly by the occurrence of lightning flashes which may cause short-period changes, of order 1000 V/cm, as they transfer charge from one part of the cloud to another. It was by measuring the changes caused by lightning in the vertical electric field at the ground, and by studying how these changes varied with the distance between the observing station and the storm, that C. T. R. Wilson first deduced how the electrical charge is distributed in thunderstorms. The underlying principle of this pioneer investigation may be demonstrated with the aid of fig. 29. We consider the thundercloud to be a bipolar electrostatic generator with positive and negative centres of charge situated approximately vertically one above the other. Lightning flashes are of two main types; cloud-to-ground discharges which generally involve the passage of charge from the lower part of the cloud to earth, and internal flashes which tend to neutralize the two charge-centres inside the cloud. Figure 29 shows the two charges A and B in the cloud and their image charges A', B' in the conducting earth. If a charge q, originally at height z_2, is lowered to earth, this involves the

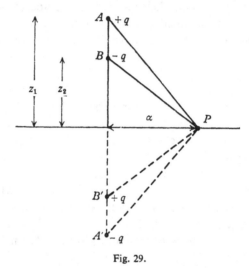

Fig. 29.

destruction of an electrical dipole BB' of moment $2qz_2$ and the vertical field-change produced at an observing station P at distance α is

$$\Delta F = \frac{2qz_2}{(z_2^2+\alpha^2)^{\frac{3}{2}}}. \tag{7.1}$$

This is the well known expression for the field along the perpendicular bisector of a dipole.

For a distant storm ($\alpha \gg z$) this reduces to

$$\Delta F \simeq \frac{2qz_2}{\alpha^3} = \frac{\mathcal{M}}{\alpha^3}, \tag{7.1a}$$

where $\mathcal{M}$ is the electric moment destroyed.

The field-change produced by an internal lightning flash involving the neutralization of A and B and therefore destruction of the dipoles AA', BB' will be

$$\Delta F = 2q \left[\frac{z_2}{(z_2^2+\alpha^2)^{\frac{3}{2}}} - \frac{z_1}{(z_1^2+\alpha^2)^{\frac{3}{2}}} \right]. \tag{7.2}$$

Again, if $\alpha \gg z_1$ and z_2, this reduces to

$$\Delta F = \frac{2q(z_2 - z_1)}{\alpha^3} = \frac{\mathscr{M}}{\alpha^3}. \tag{7.2a}$$

We see from (7.2), that for a particular value of

$$\alpha = (z_1 z_2)^{\frac{1}{3}} (z_1^{\frac{2}{3}} + z_2^{\frac{2}{3}})^{\frac{1}{2}},$$

the field-change has value zero and at this point reverses in sign. If, for example, $z_1 = 6$ km and $z_2 = 3$ km, the reversal of the field-change would occur at a distance 6·1 km from the storm.

Wilson found that the field-change produced by a cloud-to-ground discharge was positive at all distances from the storm, i.e. it tended to augment the normal fine-weather positive field, whereas the field-change due to an internal flash was positive for near flashes and negative for distant storms. This information pointed to the main positive charge of the cloud being situated above the negative charge, and revealed that most cloud-to-ground lightning flashes lower negative charge to earth. Wilson's findings have since been amply confirmed by scientists working in different parts of the world; it seems that thunderstorms are always charged positively in their upper parts and have their negative charge lower down, although the distribution of charge in individual clouds may be more complicated in detail that his simple bipolar model suggests.

The most direct information on the electrical structure of the thunderstorm comes from measurements of the magnitude and direction of the electric field made at different heights inside the cloud. The pioneer work was carried out at Kew Observatory, just before the Second World War, by Sir George Simpson and his collaborators who attached a recording apparatus, called an alti-electrograph, to a balloon. They obtained continuous records of the direction of the vertical electric field and some indication of its magnitude. The balloon also carried an aneroid barometer to give the pressure, and hence the height of the balloon up to an altitude of about 8 km, at which the alti-electrograph was released and returned by parachute. The records clearly indicated that the upper portion of a thundercloud was positively charged, the compensating negative charge being situated in the lower part of the cloud. In the majority of storms there was also evidence of one or more localized regions of positive charge in the base of the cloud, these being usually associated with heavy rain. Positive charges predominated above 7 km,

6

negative charges between 2 and 7 km, the lower positive charges being located below 2 km. The upper positive charge was usually associated with temperatures below $-20\,°C$ and always below $-10\,°C$. The temperature at the centre of negative charge was usually below $0\,°C$ while the lower positive charge was usually at temperatures above $0\,°C$. Simpson and his co-workers found that, on average, their data could be represented by a model with an upper charge of $+24$ coulombs (C) distributed in a spherical volume of 2 km radius centred at a height of 6 km ($-30\,°C$), a negative charge of -20 C in a sphere of radius 1 km centred at 3 km ($-8\,°C$), and a charge of $+4$ C in a sphere of radius 0·5 km at 1·5 km ($+2\,°C$). In other words, the bulk of the separated charges occurred in regions where the temperature was below $0\,°C$; this fact will weigh heavily when we come to discuss the origin of the charge.

   This picture of the charge distribution has been broadly confirmed by later workers using instruments carried on balloons and on aircraft, and sited on high mountains whose summits become engulfed by the clouds, but there are differences in detail. For example, recent investigations in South Africa suggest that in the large, single-cell type of storm, the negative charge is contained in a nearly vertical column, which on occasion, may be 6 km long, the temperature at its base being about $0\,°C$ and its top frequently reaching up to but not beyond the $-40\,°C$ level. On the other hand, the work of the Cambridge school suggests that this type of storm is not so common in England, where the charge and precipitation may be located in a number of small subcells, each of which may, in turn, become involved in a lightning flash.

   Information on the strength of the electric fields existing in thunderclouds is rather scanty. The alti-electrograph records rarely showed fields in excess of 100 V/cm but this was because sparking occurred at the electrodes of the instrument in stronger fields, which, no doubt, existed near the electrical centres of the storms. More recent balloon observations in the United States, using a telemetering device, have occasionally reported fields up to 2000 V/cm. Important information has also been obtained in America by Dr Ross Gunn using aircraft. The mean of the maximum field strengths encountered in each of nine storms was 1300 V/cm, and on one occasion a field of 3400 V/cm was measured just before the aeroplane was struck by lightning. The most intense fields were encountered near the $0\,°C$ level. Such fields, of order 1000 V/cm, are what we should expect to exist near the centre of a simple bipolar cloud, in which charges of order 20 C are separated by distances of some 3 km, and indicate that the total potential differences between the main charge centres may lie between 100 and 1000 million volts.

The question now arises whether, even so, these fields are sufficient to initiate a lightning discharge in the cloud. Normally, in dry air at standard atmospheric pressure, a field of about 30000 V/cm is required for breakdown with small sparks. At the pressures and temperatures existing in thunderclouds, this would be reduced to about half. However, in the presence of water drops, the breakdown field is reduced still further. Laboratory experiments show that, in strong fields, the drops become deformed and develop filaments which initiate corona discharge. With drops of 1 mm radius this phenomenon occurs in fields of 10000 V/cm; larger drops require weaker fields. Fields strong enough to *initiate* a lightning flash may occur in only small localized regions of the cloud but, once started, the discharge can be propagated in much weaker fields.

By measuring the change in the vertical component of the electric field at the ground produced by a distant lightning flash, one can use (7·1a) to calculate the electric moment $\mathcal{M} = 2QH$ destroyed, where Q is the charge neutralized and H is the vertical separation of the main positive and negative charge centres for an internal flash, or the height of the negative charge above ground for a discharge to earth. A mean value of $\mathcal{M}$ for all types of flash is 110 C.km; taking values of H ranging from 2 to 3·5 km for cloud-to-ground discharges and up to 5 km for internal flashes, the charges neutralized are of order 10–30 C. As lightning flashes from an individual thunderstorm cell occur at intervals of about 20 s, the current dissipated by lightning is of the order 1 amp.

Negative charge is generally transferred to the earth by cloud-to-ground discharges, but during internal flashes negative charge is invariably moved upwards. Another very important source of charge dissipation is the point-discharge current (St Elmo's Fire), which sets in at pointed objects, such as trees and blades of grass, whenever the electric field exceeds some critical value (usually about 10 V/cm), for the particular object. With a thunderstorm overhead the point-discharge currents usually transfer positive charge upwards tending to neutralize the negative charge in the lower part of the cloud. This current is an important agent in discharging the thunderstorm 'dynamo', which also loses positive charge from its upper pole to the highly conducting upper atmosphere above, and charges of both signs on precipitation elements. The passage of a lightning flash is accompanied by an almost instantaneous drop in the electrical field strength as measured at the ground. Almost immediately the field starts to recover, generally rapidly at first and then more slowly. The initial rate of recovery suggests that if this rate were maintained the cloud would, on average, become re-charged in about 7 s, but leakage of charge by point-discharge currents, natural ionic currents, and rain increases this time to about 20 s, after which a second flash may occur.

The rebuilding of electric moment implies not only a generation of charge but a vertical separation of charges of opposite sign. However, there is evidence to suggest that the changes of electric field represented by the recovery curves are caused, at least partly, by rearrangement of the space charge existing around and above the storm so that these may not give reliable information on the re-charging of the thunderstorm itself.

The structure of the lightning flash

The lightning flash may pass from cloud to ground or internally between the negative and positive poles within the cloud. Cloud-to-ground flashes are easily seen and photographed and have been studied more extensively. There is evidence that in clouds containing the pocket of positive charge near the base, cloud-to-ground discharges often originate in the intense electric field which exists between this positive charge and the base of the main negative column. The complex structure of a single flash has been revealed in recent years mainly by Sir Basil Schonland and his collaborators in South Africa, using a special moving camera based on an original design by C. V. Boys. In its original form, the camera had two lenses arranged to rotate at opposite ends of a diameter of a circle. Images of the lightning flash were recorded on a stationary film behind the lenses. Each lens distorted the image of a non-instantaneous flash, but the two images were distorted in opposite directions so that, from a comparison of the two pictures and a knowledge of the velocities of the lenses, it was possible to deduce the direction and speed of the visible discharge processes. In a later, improved, version of the camera the two lenses were fixed and the film moved. Photographs taken with such cameras reveal that the lightning flash may sometimes consist of a single stroke and sometimes of a number of successive strokes following the same track in space at intervals of a few hundredths of a second. The average number of strokes is about three, but as many as 14 have been recorded in a single flash.

A cloud-to-ground discharge is initiated by a streamer which develops downwards from near the base of the negatively charged column in the cloud in a series of steps, 10–200 m long and separated by time intervals varying between 15 and 100 μs. Each step is revealed in the photograph by a sudden increased luminosity of the freshly ionized air at the tip of the streamer. This *stepped leader* approaches the ground at an average speed of about 10^7 cm/s, often along a zig-zag path with downward pointing forks or branches; hence the term 'forked lightning' (see pl. XXIII). When this leader stroke reaches to within 5–50 m of the earth, a streamer from some point on the earth comes up to meet it, and then there commences the upward *main or return stroke* which travels up the ionized channel established by the leader. The luminosity of this return stroke is much greater than that of the leader; it travels at speeds of about 3×10^9 cm/s (one-tenth the velocity of light) and lasts for only about 100 μs. It is the return stroke which carries the main current of the discharge; this is typically of order 10000 amp. though currents of 100000 amp. are occasionally measured. The return-stroke channel is only a few centimetres in diameter but the core of the channel that carries most of the current may be as little as a few millimetres across. After the passage of the stepped-leader and first return stroke there may be an interval of a few hundredths of a second followed by a second leader and return stroke. The leaders to the second and subsequent strokes usually consist of streamers which develop in a single flight from cloud to ground travelling at speeds of about 2×10^8 cm/s; these are called *dart-leader strokes*. Following the dart leader there is a return stroke travelling from the ground up to the cloud. The average time interval between the component strokes of a flash is about 0·05 s, the average duration of a complete discharge being about 0·25 s.

Schonland and his colleagues, from a study of the variations in electric field which occur in the intervals between the separate strokes of a lightning flash to ground, conclude that these separate strokes are due to an intermittent discharge process which taps the negatively charged column at progressively higher and higher levels. The discharge process, which does not extend above the -40 °C level, is interpreted as one which links the positive, highly

Plate xxiii. Lightning strokes travelling along branched paths from cloud-base to ground. Some branches apparently terminate in the air without reaching the ground. (Photograph by W. J. S. Lockyer.)

ionized, branched top of the main lightning channels successively with higher regions of the negative column by upwardly directed positive *junction streamers*. The interval between strokes is thus the time required to connect previously untapped regions of negative charge with the top of the lightning channel formed by the preceding stroke. However, Brook and his collaborators, working in New Mexico, challenge this interpretation and find that in about 25% of inter-stroke intervals a continuing current flows to earth and is accompanied by a continuous luminosity that may last for 40–300 ms.

A large proportion (about two-thirds in England and nine-tenths in South Africa) of lightning discharges occur entirely within the cloud itself, and some progress towards the earth without reaching it. Internal flashes are more common in thunderstorms with high bases, when it is easier for the discharge to pass between the lower negative and upper positive poles of the cloud than from one of these to ground. The luminous structure of intra-cloud flashes is usually obscured by the intervening cloud droplets so that only a diffuse illumination (sheet lightning) is seen. Photographs show that the cloud remains luminous throughout the duration of the flash but that there are intermittent bursts of brighter luminosity that cause the flickering appearance of the internal discharges. The duration of internal discharges varies between about 150 and 500 ms. The electric field-change records indicate that the internal discharge is initiated by a slowly-descending streamer carrying positive charge and lasting for 100–300 ms followed by a more rapid return stroke of 50–200 ms duration carrying negative charge upwards. The structure of these internal discharges appears to be similar to that of the long air discharges that begin inside the cloud and end in the clear air outside and which have continuously luminous leaders, and return strokes that are much less rapid and intense than when the leader reaches the conducting earth. Field-change records of internal discharges reveal the return stroke to be composed of several rapid components of 1–3 ms duration each carrying currents of 1000–4000 A.

About three-quarters of the energy released in a lightning flash, which amounts to about 10^{10} joules, is spent in heating up the narrow air column surrounding the discharge channel, the temperature rising in a few microseconds to about 30000 °C. The air in the channel expands explosively to create intense sound waves which are heard as thunder.

Correlation between lightning and precipitation

Lightning is usually accompanied by heavy precipitation although, in warm, dry climates, this may not reach the ground. Most theories of thunderstorm electrification have assumed that precipitation plays an important role in the generation and separation of the electric charge and, as the main charge centres appear at levels where the temperature is below 0 °C, there has been a tendency to associate that generation with the presence of supercooled water and/or the ices phase. Certainly lightning often accompanies heavy showers composed of small hail pellets sometimes mixed with raindrops. Kuettner, from observations made in thunderstorms capping the Zugspitze in Germany, reported that solid precipitation elements were dominant in the greater part of the thundercloud and were present on 93 % of occasions. Snow pellets and pellets of soft hail were the most frequent forms of hydrometeor, being present on 75 % of occasions, and were always accompanied by strong electric fields, but large hail was relatively rare.

Fitzgerald and Byers of the University of Chicago, using aircraft fitted with electric-field meters, reported that actively building regions of thunderstorms were regions of excess negative charge and that the electric fields, which increased rapidly with the onset of precipitation, indicated that the initial precipitation streamers carried a negative charge towards the earth, leaving a positive space charge in the upper regions of the cloud. The strongest fields, of up to 2300 V/cm, were associated with regions of heavy precipitation. In particular, a large hail shaft produced a strong, smoothly increasing field that indicated a negative charge on the hail. Malan and Schonland found that, in South African storms, the negative charge is often distributed in a nearly vertical column which may extend up to, but not beyond, the −40 °C level. This is consistent with the charge being generated by growing hail pellets, because supercooled droplets exist at temperatures down to, but not below, −40 °C.

Reynolds and Brook, making simultaneous observations with a 3 cm radar and measurements of the electric field on Mt Withington in New Mexico, reported that intensification of the electric field and the radar echo went hand in hand. The initial radar echo usually appeared at about the −10° C level and the time interval between this and the first lightning flash was about 12 min. On many days, when clouds of considerable depth and vigour developed without precipitation, the electric field showed no significant departure from the fine-weather value.

On the other hand, Vonnegut and Moore, using the same radar equipment in the same location, claimed that, in some of the New Mexico storms, electrification of the clouds and even lightning began before the arrival of precipitation at the mountain top, and when the precipitation rate inside the cloud, as deduced by radar, was only a few millimetres per hour and occasionally 1 mm/h. However, heavy gushes of rain or hail often reached the ground 2 to 3 min after a lightning flash, and Vonnegut and Moore take this as evidence that the lightning is the cause rather than the result of the precipitation.

It is, in fact, often difficult to correlate, in both space and time, electrical activity and lightning with the development of precipitation as seen by a single radar. Lightning flashes often take a rather long and tortuous path, and it is almost impossible to locate them accurately in space, especially in the daytime, and to identify them with a particular precipitation echo on the radar screen. The evolution of the radar echo itself is also open to misinterpretation, particularly at short ranges for which the receiver may be insensitive, and when the storm is growing in a strong vertical wind shear. Until much more precise observations have been made, and bearing in mind that it will usually take 2 or 3 min for the rain to fall from cloud base to ground, the available evidence may be interpreted as showing that the onset of lightning and heavy precipitation *within* the cloud are practically simultaneous. The appearance of heavy gushes of precipitation at the ground a few minutes after a lightning flash is most readily explained by the charged hydrometeors being suspended in the strong electric field until this has been partly destroyed by the flash, the reduction in their falling speed being proportional to the square of the field strength (see (7·8)).

Vonnegut and Moore challenge the generally accepted thesis that precipitation is the main source of electrification on the additional grounds that the charge on thunderstorm rain reaching the ground is usually smaller in magnitude, and often of opposite polarity, to what one would expect if the principal negative charge of the storm were carried down by the precipitation. However, the charge on the rain reaching the ground has probably been acquired largely by the capture of positive ions produced by point-discharge at the Earth's surface by the thunderstorm field, and so any original negative charge carried by hail pellets may be neutralized, and even reversed, during

7

their melting and fall through the positive space-charge blanket towards the ground.

Although the main charge centres of large thunderstorms appear always to be located in the subfreezing part of the cloud, there are on record a few reports of lightning being observed from clouds whose tops were warmer than 0 °C and could therefore have contained no ice. Unfortunately, these observations, some made from aircraft flying over the sea in twilight or darkness, are not convincing. However, should detailed and well-documented observations establish that lightning occurs in non-freezing clouds, however infrequently, it will be necessary to look for a mechanism not involving the ice phase.

Basic requirements of a satisfactory theory

The main features of the thunderstorm, which have been described in the foregoing paragraphs, and with which any satisfactory theory of charge generation and separation must be consistent, are as follows:

(i) The average duration of precipitation and electrical activity from a single thunderstorm cell is about 30 min.

(ii) The average electric moment destroyed in a lightning flash is about 100 C km, the corresponding charge being 20 to 30 C.

(iii) In a large, extensive cumulonimbus, this charge is generated and separated in a volume bounded by the -5 and -40 °C levels and having an average radius of perhaps 2 km.

(iv) The negative charge is centred near the -5 °C isotherm, while the main positive charge is situated some kilometres higher up; a subsidiary positive charge may also exist near the cloud base, being centred at or below the 0 °C level.

(v) The charge generation and separation processes are closely associated with the development of precipitation, probably in the form of soft hail.

(vi) Sufficient charge must be generated and separated to supply the first lighting flash within 10 to 20 min of the appearance of precipitation particles of radar-detectable size, and to establish large-scale electric fields of at least a few kilovolts per centimetre.

The search for a new theory of thunderstorm electrification

In a critical review, published in 1953, of then current theories of thunderstorm electrification, the author concluded that none was capable of generating and separating charge at a sufficient rate to account for the observed frequency of lightning flashes and the changes of field accompanying them. In particular, the 'influence' theories of Wilson, published in 1929, and of Frenkel, published in 1947, based on the selective capture of negative ions by raindrops polarized initially by the positive fine-weather electric field, are incapable of producing large-scale fields stronger than about 500 V/ cm because, in stronger fields, the drift velocities of the ions would exceed the terminal velocities of the raindrops and so prevent continued ion capture. Simpson's drop-breaking mechanism (see p. 175) also appeared to be much too weak and, in any case was of the wrong polarity. An ingenious mechanism, suggested by Workman and Reynolds in 1948 and involving the shedding of water drops from the surfaces of wet hailstones, not only made the presence of *large* hail mandatory, but appeared so sensitive to the concentrations of dissolved salts, and gases such as carbon dioxide and ammonia, that it was deemed unattractive as a universal mechanism of charge generation in thunderstorms.

In searching for a more satisfactory explanation, the author was impressed by the following pieces of evidence:

(i) The primary charge mechanism appears to operate between the 0 and $-40\,°C$ levels and is therefore likely to be associated with supercooled water drops and/or ice particles.

(ii) Lightning frequently accompanies soft hail, which is produced by the accretion and freezing of supercooled cloud droplets to form rime-like aggregates.

(iii) Several workers had reported that during the formation of rime the deposit acquired a negative charge, though Findeisen, working in Prague during the Second World War, obtained positive charging and assumed that the compensating negative charge was carried away on small splinters of ice ejected from the rime surface.

Using the rates of charging recorded in these experiments, and assuming that similar rates might apply during the growth of hail pellets by riming in the supercooled parts of the cloud, Mason demonstrated that this might provide a powerful mechanism of thunderstorm electrification. However, there were a number of doubts conconcerning the experiments themselves. In particular, little attention was paid to the possibilities that the observed electrification may have been caused by charges on the incident droplets, or by splashing rather than freezing of droplets on the ice surface. Splashing may well have produced the positive charging observed by Findeisen.

Moreover, Reynolds and his colleagues, working in New Mexico, reported that simulated hailstones, made by rotating an ice-coated metal sphere in a cloud of supercooled droplets, showed no detectable charging unless the cloud also contained ice crystals.

This led Latham and Mason in 1960 to make an extensive study of the electrification produced by the impact and freezing of supercooled droplets on a simulated hailstone, with the intention of eliminating spurious charging effects, and of investigating the dependence on the size, impact velocity and rate of freezing of the drops. They found that, provided the droplets did not splash on impact, the 'hailstone' acquired a negative charge roughly proportional to the flux of droplets, and they collected small ice particles downwind of the target that were assumed to have been ejected from its surface. In a typical experiment, with the air temperature at $-15\ ^{\circ}$C and droplets impacting at 10 m/s, 10^4 droplets of radius 40 μ struck the hailstone within 10 s and produced a total charge of 4×10^{-2} e.s.u. (13 pC), i.e. an average of 4×10^{-6} e.s.u. (1·3 fC) per drop. Droplets of $r < 15\ \mu$ produced very little charge, and when the combination of droplet size, impact velocity, and surface temperature of the hailstone was such as to produce splashing, the target acquired a *positive* charge.

On the assumption that the results of these experiments could be applied to the growth of soft hail pellets in a natural cloud, Mason was able to demonstrate that such a mechanism, operating for several minutes in a cell of modest dimensions with moderate rates of precipitation, could account for the generation of sufficient charge to supply several lightning flashes, and that gravitational separation of the negatively charged hail pellets and positively charged ice splinters could account for the observed sign and magnitude of the large-scale electric fields.

The origin of the charge liberated during the splintering of freezing drops was attributed to the newly discovered thermoelectric effect in ice by which the hydrogen and hydroxyl ions, formed by the dissociation of a small fraction of the ice molecules, become separated under the influence of a temperature gradient. The process depends on the fact that the concentrations of both ions increase quite rapidly with increasing temperature and that the hydrogen ion (proton) diffuses much more rapidly through the ice lattice than does the hydroxyl ion. Thus if a steady temperature gradient is maintained across a piece of ice, the warmer end will possess initially higher concentrations of both ions, but the more rapid diffusion of the hydrogen ions down this concentration gradient will lead to a separation of charge with a net excess of positive charge in the colder part of the ice. The space charge created in the ice by this differential diffusion will set up an internal field opposing further separation of charge and a steady state will be reached in which the net flow of current is zero. Mason calculated the potential gradient developed across a piece of pure ice by a steady temperature gradient to be

$$-dV/dT = 1\cdot9 \text{ mV/degC}, \qquad (7\cdot3)$$

and the charge transfer between two pieces of ice of different temperatures, T_1, T_2, in temporary contact, to reach a maximum value of

$$q = 3 \times 10^{-3}A(T_1-T_2) \text{ e.s.u./cm}^2 = A(T_1-T_2) \text{ pC/cm}^2 \qquad (7\cdot4)$$

after about $0\cdot01$ s, A being the effective area of contact.

The result expressed in (7·3) was confirmed by Latham and Mason, who obtained a thermoelectric power of -2 mV/degC for pure ice at temperatures between -7 and -20 °C and, even more convincingly by Brownscombe and Mason, who used an induction method to avoid the difficulties that arise in attaching metal electrodes to ice, and obtained a value of $-2\cdot3$ mV/degC at -20 °C.

Returning now to the freezing droplets, we recognize that a shell of ice forms around the outside of the drop soon after nucleation but that the rate of freezing of the liquid interior is limited by the rate at which the latent heat of fusion is dissipated by conduction and convection. During this process, a radial temperature gradient is established across the ice shell, the inner surface being held at 0 °C by the liquid water and the outer surface cooling towards the lower air temperature. According to the above theory, protons will migrate preferentially down the temperature gradient and produce a positive space charge in the outer layers of ice, so that when the drop eventually bursts by expansion of its freezing interior, ice splinters ejected from the surface will tend to carry away positive charge and leave the remainder of the drop negatively charged. A simple calculation indicates that for drops of radius 40 μ to produce an average charge of 4×10^{-6} e.s.u. (1·3 fC) in air at -15 °C, as reported by Latham and Mason, about one-tenth of their surface area would have to be removed as splinters.

Such high rates of splinter production have not been observed and, if operative during the growth of natural hail particles, would lead to rapid glaciation of the cloud and the disappearance of all the super-cooled water. It appears then, that the rates of charging reported by Latham and Mason, and also the charges of order 10^{-3} e.s.u. produced during the freezing and fragmentation of millimetre-size drops, must involve a mechanism more powerful than the thermoelectric effect but this still remains to be identified.

Although the rates of charging measured by Latham and Mason were in reasonable agreement with some of the earlier riming experiments, they have not been reproduced in some very recent experiments carried out in the Meteorological Office by Aufdermaur and Johnson. They find that a rime pellet, grown by the accretion of supercooled droplets ranging from 10 to 50 μm in radius, does not acquire a detectable charge in the absence of an external electric field, but only when droplets rebound from the surface of the pellet in the presence of an electric field applied parallel to the airstream carrying the droplets. (See p. 172 for further details.)

The possibility that hail particles, initially polarized in the positive fine-weather electric field, might acquire a net negative charge by ice crystals rebounding and carrying away positive charge from their undersurfaces, was investigated theoretically by Latham and Mason following an earlier suggestion by Muller–Hillebrand that such a mechanism could lead to a rapid exponential growth of the electric field to about 3000 V/cm above which it would be limited by the onset of corona discharge at the ground, and made calculations to show that such a mechanism could produce and separate charge at the required rate in thunderstorms, provided that the time of contact between the ice crystals and the hail pellets exceeded, or was at least comparable with, the relaxation time for conduction of charge between them.

In experiments designed to simulate this mechanism in a wind tunnel, Latham and Mason found that when ice crystals of about 50 μ diameter were allowed to impact at velocities of several metres per second on a smooth ice sphere they acquired very small charges, averaging only about 5×10^{-9} e.s.u. ($\approx$ 2 aC) per collision, which, since they increased linearly with the difference between the temperature of the target surface and that of the air and reversed in sign when this temperature difference was reversed, were attributed to the thermoelectric effect.

However, no additional charging was detected when electric fields of up to 700 V/cm were applied parallel to the airstream, probably because these small, regular ice crystals remained in contact with the smooth ice target for periods considerably shorter than the relaxation time for conduction in pure ice, which is about 10 ms at −10 °C.

On the other hand, Scott and Levin of the University of Washington have recently obtained a positive result using much larger natural snow and ice particles of diameter $\frac{1}{4}$ to $2\frac{1}{2}$ mm impacting at 1 m/s on an ice sphere placed in polarizing fields of up to only 100 V/cm. The crystals were allowed to fall into an earthed vertical metal tube, pass through an induction ring to measure their initial charge, and then to impact on the ice sphere placed between two horizontal electrodes. The sphere was connected to a sensitive electrometer capable of recording the pulses of charge produced by individual crystal impacts. These charges, when corrected for the initial charges on the ice crystals, were strongly correlated with both the magnitude and direction of the applied field, indicating that they were acquired by the ice sphere as the result of the rebounding ice crystals carrying away some of its polarized charge. The actual magnitudes of the average charges carried away by crystals of an average radius r were in fair agreement with the theoretical value, $\frac{1}{2}\pi^2 Fr^2$, at least for fields of up to 50 V/cm; but with some indication that the measured charges were greater than the theoretical values in stronger fields of up to 100 V/cm. The fact that Scott and Levin observed this type of charging, while Latham and Mason did not, may be attributed to the conductivity of their natural snow crystals being probably at least an order of magnitude higher than that of Latham and Mason's pure ice crystals; moreover, by virtue of their rough surfaces and low impact velocities, they probably remained in contact with the ice sphere for considerably longer times.

On the evidence of these experiments and the calculations mentioned above, it seems highly likely that collisions between ice crystals and polarized hail pellets provide a powerful mechanism of thunderstorm electrification, in which the field builds up rapidly through a positive feedback mechanism until it is checked by electrical forces counteracting gravitational separation of the particles, by leakage currents and, finally, by the onset of lightning. It may be enhanced, and even overshadowed by a similar mechanism in which the polarized charges on the hail pellets are also removed by the rebound of a small fraction of the colliding cloud droplets. This latter process might well be favoured by the fact that it will not be limited by the relaxation time for conduction which, for liquid water, is only a few microseconds.

We shall now formulate the theory of these two mechanisms in more detail and more realistically than hitherto, and calculate the rates at which they might generate electric charges and fields in a simple model of a thunderstorm in which the precipitation is allowed to build up realistically from very small values, and in which account is taken of the fact that separation of the charged particles will become increasingly retarded because of the forces exerted on them by the growing field.

Electrification produced by the collision of cloud particles with polarized hail pellets

A hail pellet falling in a downwardly directed electric field may be considered as a conducting sphere that becomes polarized with its upper half negatively charged and its lower half positively charged —see fig. 30. Cloud particles (droplets or ice crystals) colliding with, and rebounding from, the lower half will carry away positive charge and leave a negative charge on the pellet, provided that the time of contact exceeds the time required for charge transfer by conduction between the two particles. The rate of charging of the hydrometeor of radius R, falling at velocity V_R relative to the air and overtaking the smaller cloud particles of fall velocity v_r, radius r and number concentration n_r, in a vertical field F, is given in the c.g.s./e.s.u. system as

$$\frac{dQ_R}{dt} = -E'\pi R^2(V_R-v_r)\, n_r \alpha r^2 \left(\frac{\pi^2 F}{2}\cos\theta + \frac{\pi^2}{6}\frac{Q_R}{R^2}\right)$$

$$= \frac{1}{\tau}(3FR^2\cos\theta + Q_R), \tag{7.5}$$

where E' is the collision cross-section, i.e. the fraction of cloud particles lying in the cylinder swept out by the hydrometeor which actually *collide* with it, α is the fraction of these particles which *rebound* from it or make grazing contact with it, and θ is the angle between the field and the line of centres of the two colliding particles (fig. 30).

Now $\tau = (\frac{1}{6}\pi^3 E'(V_R-v_r)n_r\alpha r^2)^{-1}$ is the relaxation time for charging of the hail pellet and, although it is a function of R because of its

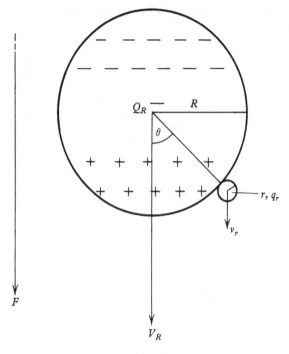

Fig. 30

dependence on V_R, we shall treat it as a constant and investigate the effect of changing its value.

In the absence of experimental evidence to the contrary, we shall assume that ice crystals rebound with equal probability from all parts of the undersurface of the hail pellet and adopt an average value for $\bar{\theta} = 45°$. With these two simplifying assumptions, (7.5) may be integrated to give

$$Q_{R(t)} = \frac{-3}{\sqrt{2}} \frac{1}{\tau(e^{t/\tau}, -1)} \int_0^t R^2(t) F(t) \, dt. \tag{7·6}$$

Gravitational settling of the negatively charged hail and the rebounding positively charged cloud particles will intensify the electric field at a rate given by

$$\frac{dF}{dt} = -4\pi[\overset{R}{\Sigma} N_R V_R Q_R - \overset{r}{\Sigma} n_r v_r q_r - 10^3(e^{F/5} - 1)], \tag{7·7}$$

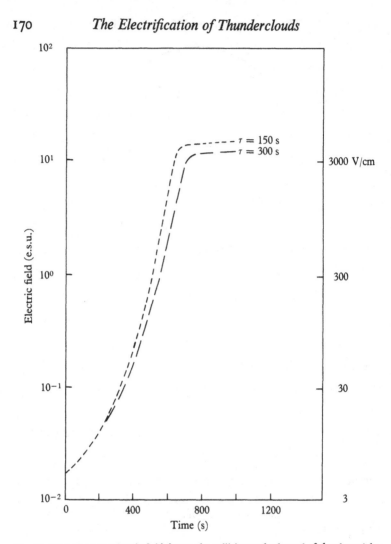

Fig. 31. Growth of the electric field due to the collision and rebound of cloud particles and polarized hail pellets. (From Mason, *Quart. J. Roy. Met. Soc.* **99** (1973), 398.)

where N_R, V_R are respectively the number concentration and falling speed relative to the air of hail pellets, $n_r v_r q_r$ refer to the rebounding crystals, and the last term is an empirical expression representing the leakage of current due to point discharge beneath the cloud. V_R is determined by the action of gravitational and electrical forces acting on the hail pellets, *viz.*

$$V_R \simeq \tfrac{20}{3} R\rho g - \frac{5FQ_R}{\pi R^2} = V_R' - \frac{5FQ_R}{\pi R^2}, \qquad (7\cdot8)$$

where ρ is the density of the hail pellet and V_R is the fall speed in the absence of an electric field.

Now N_R, R, and V_R' are related to the precipitation intensity in the absence of an electric field, expressed in terms of an equivalent rate of rainfall, by

$$p = \tfrac{4}{3} \pi\rho \overset{R}{\Sigma} N_R R^3 V_R', \qquad (7\cdot9)$$

and if the hail pellets are assumed to grow by accretion in a cloud of constant liquid-water concentration, p will grow exponentially with time according to

$$p = p_0 e^{t/T}, \qquad (7\cdot10)$$

T being the appropriate time constant.

Ignoring for the moment the subsidiary $n_r v_r q_r$ term (which how-ever is included in the calculation of fig. 31), the five equations $(7\cdot6)$–$(7\cdot10)$ contain five interdependent variables, i.e. Q_R, R, V_R, p, F, all of which can be calculated as functions of time if τ, T, and initial values of p and F are specified and use is made of the following empirical relationships between weighted mean values of R and V_R' and p, viz.: $\overline{R} = \tfrac{1}{20} \rho^{-\frac{1}{3}} p^{\frac{1}{4}}$ and $V_R' = \tfrac{1}{3}\rho^{\frac{2}{3}} p^{\frac{1}{4}}$ with R measured in cm, V_R' in cm/s and p in mm/h.

Fig. 31 shows the results of a computer calculation of the growth of the electric field from an initial value of 5 V/cm, given $p_0 = 1$ mm/h, $T = 150$ s, $n_r = 50$/litre, $r = 50\,\mu$ and $\alpha = 1$ for ice crystals. The curve for $\tau = 150$ s shows the field to grow slowly at first to reach 70 V/cm after 400 s but then to build rapidly to reach a maximum value of 4200 V/cm only 5 min later. At this stage, when the precipitation intensity is 106 mm/h and the total rainfall up to this time is 4·4 mm, the gravitational forces on the hail pellets are effectively counteracted by those due to the electric field.

The effect of changing τ from 150 to 300 s is shown by the second curve of fig. 31.

Charging produced by cloud droplets rebounding from hail pellets

Although a high proportion of the supercooled droplets colliding with a rimed hail pellet are likely to freeze on contact and contribute to the growth of the pellet, it is quite possible that a small fraction will rebound or splash off, and carry away some of the induced charge. We shall now calculate what fraction of droplets would have to rebound in order to build up a strong electric field within about 10 min and examine the experimental evidence for such a mechanism.

The theory for the growth of the field is almost identical to that derived above for collisions between hail particles and ice crystals, the flux of rebounding droplets entering only through the factor $\tau = (\frac{1}{6}\pi^3 E'(V_R - v_r) n_r \alpha r^2)^{-1}$, the value of which is not critical. In the present context n_r is the number concentration of cloud droplets of radius r and α is the fraction of colliding droplets that rebound. If, for example, we put $\overline{(V_R - v_r)} = 500$ cm/s, and now take $r = 10 \mu$, we arrive at a value of $\tau = 300$ s for $\alpha n_r = 1/\text{cm}^3$. Since $n_r \simeq 100/\text{cm}^3$, this implies a requirement for only about 1% of the impacting droplets to rebound from the hail pellets in order to build up electric fields at the rate shown in fig. 31.

An experiment to explore these possibilities has recently been carried out in the Meteorological Office by Aufdermaur and Johnson.[*] They have investigated the charging of an ice pellet formed by the accretion of supercooled droplets in the presence of a polarizing electric field, using ingenious electronic techniques to measure, correlate and display the charges produced by the impact of individual droplets and to exclude any spurious charging events from the analysis. The ice pellet, supported in a small vertical windtunnel contained in a cold room, was grown by the accretion of droplets of distilled, deionized water whose diameters could be selected in the range 20 to 100 μ. Irradiation with a β-ray source reduced the average charge on the incident droplets to $< 10^{-7}$ e.s.u. (≈ 0.03 fC). The experiments were conducted with air temperatures between -5 and -15 °C and with an air speed of 10 m/s. The ice pellet was held on an insulating support between two electrodes which produced electric fields of up to 1500 V/cm at the surface of the pellet. An upper induction ring detected charged droplets or other particles entering the apparatus and excluded them from analysis by rendering the measuring circuits insensitive until the particle had

[*] Aufdermaur and Johnson, *Quart. J. Roy. Met. Soc.*, **98** (1972), 369.

TABLE 10. *Charges on droplets rebounding from a polarized rime pellet*

(After Aufdermaur and Johnson, *loc. cit.*)

Droplet radius (μ)	F (V/cm)	q (fC) Theor.	Obs.
10	1500	3	6
25	1500	17	30
50	500	23	18

passed through the apparatus. A genuine charging event produced charges of equal magnitude and opposite sign on the pellet and on the droplet leaving it, the latter being recorded by a lower induction ring. Both these charges were measured by electrometers having a sensitivity of 6×10^{-6} e.s.u. (2 fC), correlated, displayed on a storage cathode-ray tube and recorded on magnetic tape for subsequent analysis.

The experiments showed that the growing rime pellet did not acquire a detectable charge in the absence of an applied electric field but, in the presence of a field, significant charging events, corresponding in number to between 0·1 and 1·0 % of the number of accreted droplets, were recorded. The ice pellet acquired charge of the same sign as the potential of the upstream electrode in conformity with the induced charge on its upper surface being removed by the rebound of some of the incident droplets. The fact that some of the droplets caught downstream of the target were smaller than the incident droplets was taken as evidence that the rebounding droplets left some of their mass behind on the pellet, where they, in effect, experienced 'partial coalescence'. Table 10 shows some representative measurements of the average charges produced for various values of the applied field and droplet size, and compares these with the estimated theoretical values.

Bearing in mind that, because of the experimental difficulties, the measured charges had a probable error of about 50%, the ice pellet was originally cylindrical rather than spherical, that its geometry changed considerably during growth, and that the electric field probably varied considerably over the surface of the target, the agreement between observed and theroretical values is as good as can be expected and good enough to establish the nature of the charging process.

The experiments and calculations described in this section appear to have demonstrated that collisions by droplets are likely to be at least as effective as the rebound of ice crystals in charging polarized hail pellets and therefore in producing strong electric fields in thunderstorms. In the absence of experiments to demonstrate con- clusively that small ice or snow crystals, impacting at *several metres per second* on rime pellets, can remain in contact for times sufficient to allow conduction of charge between them, present evidence favours the droplet mechanism. However, both processes may well act simultaneously and additively in the cloud and offer the most con- vincing explanation so far for the primary electrification of the thunderstorm.

The whole process is initiated by the formation of hail pellets and their becoming polarized in the weak, positive, 'fine-weather' field. The mechanism of charge generation and separation is self-accelerating at first, but later becomes self-limiting as the electrical forces on the charged particles and the leakage currents grow and oppose the primary charging current. It appears capable of producing large- scale fields of up to about 4000 V/cm within 10 min of the start of precipitation and (see below), of separating sufficient charge to supply lightning flashes at the required rate.

Magnitude of the separated charge and recovery of the field between flashes

The magnitude of the charge generated and separated by gravita- tional separation of the negatively charged hail pellets and the rebounding positively charged cloud particles is given by

$$Q_s = AF/4\pi, \tag{7·11}$$

where F is the field strength and A the cross-sectional area of the charge-generating region of the cloud. Taking the maximum large- scale field to be 4200 V/cm (14 e.s.u) reached in the model calculation of fig. 31 after 700 s, when the precipitation intensity is 106 mm/h, the separated charge in a small cell of radius 1 km would be 12 C, 48 C in a modest cell of radius 2 km, and 300 C in a very large cell of radius 5 km.

If, in the first case, a lightning flash were to neutralize 10 C of the 12 C of separated charge, the field would fall almost instantaneously to 700 V/cm and then start to recover. In the modest-sized cell, a lightning discharge of 20 C would cause the field to drop to 2450 V/ cm, while a flash of 30 C in the very large cell would reduce the field only slightly to 3800 V/cm. If the precipitation intensity were to remain steady at its pre-discharge value, the field would recover to about 4000 V/cm quite rapidly—in the case of the modest cell within about 20 s—which is about the average time interval between flashes from a cell of this size. The magnitude and frequency of lightning flashes may be expected to increase as both the size of the storm and the intensity of the precipitation increase. However, because the inductive charging mechanisms are self-limiting and cannot produce large-scale fields much in excess of 4000 V cm, the initiation of the lightning discharge may depend either on localized regions of stronger field, or on streamers being initiated at the surface of precipitation elements as described on page 180. The vertical separation of the positive and negative charges in our model cloud is

$$z = \int_0^t (V_R - v_r)\, dt$$

and the electric moment of the cloud just before the first discharge by $\mathcal{M} = 2Q_s z$. For the calculation represented by fig. 31, with τ 150 s, $t = 700$ s, $p = e^{t/150}$ mm/h, z turns out to be 2 km. The electric moment of our modest cloud would then be 200 C km and a lightning discharge neutralizing 20 C of charge would destroy a moment of 80 C km, in good agreement with observation.

Electrification produced by the breakup of large drops

(i) Experimental and theoretical studies

That the breakup of large water drops in air is accompanied by a separation of charge was first established by Simpson in 1909. Drops disrupted in a strong vertical air jet produced several large fragments carrying positive charges, while the surrounding air acquired a negative charge. Rupture drops of distilled water, about 8 mm in diameter, produced on average a charge of 5.5×10^{-3} e.s.u. (1.8 pC) or about 2.3×10^{-3} e.s.u. (or 0.8 pC) per cm^3 of water. This result has been confirmed since by several workers with some evidence that the charging depends markedly on the violence with which the drops are shattered.

Large drops breaking up in free fall, or in a steady upcurrent, first become flattened and then the lower surface becomes concave as the hydrodynamic forces overwhelm the surface tension and hydrostatic

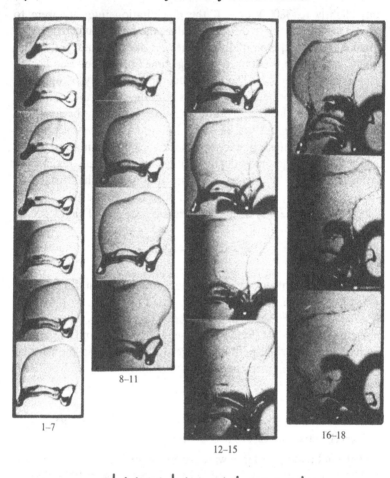

1–7

8–11

12–15

16–18

0   15 cm

Plate XXIV. A large falling water drop becomes unstable and forms an expanding bag supported on a toroidal ring of liquid. The bag eventually bursts producing large numbers of small droplets and the toroid breaks up into several large drops. The photographs were taken at intervals of 1 ms. (From Mathews and Mason, *Quart. J. Roy. Met. Soc.* **90** (1964), 275.)

forces. The surface layers of the drop are sheared off to form a rapidly expanding bag attached to an annular ring or toroid that contains the bulk of the water. Eventually the bag bursts to form a fine spray and the toroid breaks up into a circlet of much larger drops, as shown in pl. XXIV. The formation of the bag, which takes

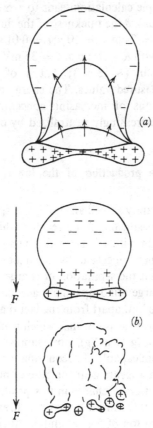

Fig. 32. Separation of charge during the breakup of a large raindrop. (a) Bursting drop. (b) Drop breaking in external field.

about 10 ms, is caused by rapid acceleration of the liquid from the surface of the toroid. This involves shearing of the electrical double layer that exists at a water–air interface as a consequence of water being composed of molecular dipoles, and the transfer of negative charge to the bag, a compensating positive charge being retained by the toroid—see fig. 32(a).

In the case of pure water, an approximate expression for the maximum charge acquired by the toroid of radius R is:

$$Q_{max} = DR\zeta\tau t/2\eta, \qquad (7.12)$$

where ζ is the electrokinetic potential of the air–water interface, τ the shear stress on the expanding liquid surface, and t the time taken to form the bag. A simple calculation leads to $\tau \simeq \rho L \delta / 2t^2$, where ρ is the density of the liquid, δ the thickness of the film, and L the height of the bag. Putting $L = 2$ cm, $\delta = 10\,\mu$, $t = 0.01$ s, gives $\tau \simeq 10$ dyn/cm^2 which, together with $R = 1.5$ cm, $\zeta = 35$ mV $= 1.2 \times 10^{-4}$ e.s.u., gives $Q_{max} = 0.04$ e.s.u. (~ 13 pC); that is, of the same order of magnitude as the measured values. The charge may be expected to decrease with solutions of increasing concentration because the separated charge is increasingly neutralized by conduction through the liquid film.

(ii) Relevance to the production of the lower positive charge in thunderclouds

Falling raindrops rarely exceed an equivalent spherical diameter of 6 mm and it may be assumed that, on reaching this size, they become unstable and break up with the large fragments acquiring positive charges in the manner suggested by the laboratory experiments. Simpson proposed that the separation of charge associated with the repeated rupture of large raindrops might account for the electrification of thunderstorms but, apart from the fact that this would charge the cloud in the opposite sense to that which is observed, Simpson's value of 2×10^{-2} e.s.u./g (7 pC/g), now supported by several later workers, does not seem capable of accounting for even the subsidiary positive charge which appears near the base of most thunderstorms. The spatial concentration of charge in this positive pocket has been estimated at about 10 C km^3, whereas the breaking of raindrops present in a concentration of 5 g/m^3 and producing 2×10^{-2} e.s.u./g (7 pC/g) would yield only 3×10^{-2} C/km^3. However, in a thunderstorm, raindrops will acquire induced polarized charges of order FR^2 in the presence of the strong primary electric field—see fig. 32(b), and the disruption may therefore result in the separation of much larger charges than were measured in the laboratory experiments just described. Thus a field of only 1.5 kV/cm (5 e.s.u.) should induce charges of about 0.5 e.s.u. (150 pC) on drops of radius 3 mm, equivalent to 4 e.s.u./g (1300 pC/g).

These considerations led Matthews and Mason in 1964 to study the breakup and electrification of large drops falling freely in still air, and in electric fields of up to 1.5 kV/cm. Drops of diameter about 12 mm, after falling 12 m, broke up between a pair of horizontal electrodes producing a vertical field. The charges on the larger fragments were measured as they fell through induction cylinders without splashing, and the volume of the fragments determined by collecting and weighing them. The mode of breakup was studied by means of high-speed photographs taken at the rate of 2000/s (see pl. XXIV). In the absence of an applied field the breaking drops produced charges of about 10^{-2} e.s.u./g (3 pC/g) in agreement with earlier workers, but the average charge per unit mass of water increased with increasing field strength and reached 5·5 e.s.u./g (1800 pC/g) in fields of 1·5 kV/cm.

These charges are two orders of magnitude greater than those quoted by Simpson and others, and suggest that large raindrops breaking in the presence of the primary field of the thunderstorm may contribute significantly to the charge configuration in the lower part of the cloud. Drops breaking well above the 0 °C level will be subjected to the positive primary field and so acquire a negative induced charge on the bag and a positive charge on the toroid, but those breaking near the cloud base, i.e. below the storm's main negative charge, will carry induced charges of opposite polarity. This may account, at least in part, for the fact that although the heavy rain falling from the central part of a thunderstorm is predominantly positively charged, an appreciable fraction of the individual drops carry net negative charges. Other negatively charged drops may arise from the melting of hail pellets which, as we have argued, acquire a negative charge in the primary thunderstorm field.

However, there is evidence to suggest that any charges acquired within the cloud above the 0 °C level are likely to be masked by the capture of positive ions produced by point discharge and drop splashing at the ground, and which constitute an upward moving current into the cloud base. The pocket of positive charge, situated in the base of the cloud and associated with heavy rain, probably owes its existence to ion capture, and the breakup of raindrops is thought to make only a secondary contribution.

The role of the lower positive charge

The fact that the presence of the lower positive charge often distinguishes a thunderstorm from a heavy shower, suggests that it may play an important role in initiating or triggering lightning discharges. Since the large-scale electric fields measured in thunderstorms do not exceed a few thousand volts/cm, and are necessarily limited to such values if they are created by the charging of hydrometeors well below 1 cm in diameter, they are much less than those required to initiate a discharge in dry air. We may conclude, therefore, either that electrical breakdown occurs much more easily in the thunderstorm in the presence of hydrometeors and/or that, if strong fields are required, they must exist in rather localized regions of the cloud that are likely to escape detection by balloons or aircraft. It seems likely that a number of such localized regions may exist in storms in which the distribution of precipitation and charge are often much more complex than the simple bipolar model suggests; but the region between the lower positive charge and the base of the main negative column is an obvious candidate. Moreover, Malan and Schonland obtained evidence that a high proportion of the stepped-leader strokes of cloud-to-ground lightning flashes originate in just this region.

Furthermore, Macky found that water drops became elongated in the direction of a strong external electric field and when this exceeded a critical value given by $F_c R^{\frac{1}{2}} = 13$ e.s.u., where R is the drop radius in centimetres, the drops became unstable, and streamers developed at the ends with the onset of corona discharge. Macky's result is in good agreement with the theoretical criterion for the onset of hydrodynamic instability, namely: $F_c(R\gamma)^{\frac{1}{2}} = 1\cdot625$, where γ is the surface tension of the liquid. Abbas and Latham found that the critical field for disintegration of a water drop also depends on the magnitude and sign of the net charge Q carried by the drop and give $F_c R^{\frac{1}{2}} = 13\cdot6 - 8Q$, where F and Q are expressed in e.s.u. Thus a drop disrupts and produces streamers more readily if it carries a charge of the same sign as the polarizing field; for example, while an uncharged drop of radius 2 mm would disintegrate in a vertical field of 9150 V/cm, a similar drop carrying a reinforcing charge of 0·2 e.s.u. (66 pC) would require a field of only 8·1 kV/cm. These fields are much lower than the 30 kV/cm required to initiate a breakdown between metal electrodes in air at full atmospheric pressure, though still rather greater than the large-scale fields encountered in thunderstorms.

We therefore envisage that, in the lower part of the cloud, below the 0 °C isotherm, the falling raindrops become positively charged, mainly by the capture of positive ions produced near the ground by point discharge and the splashing of raindrops under the influence of the negative primary thunderstorm field. This produces a concentration of positive charge in the base of the cloud and, in the strong negative field that will exist between this and the base of the main negative column, raindrops, produced largely by the melting of negatively charged hail pellets, will become polarized with their upper halves positively charged and their lower halves negatively charged. If the local field is strong enough to induce hydrodynamic instability followed by corona discharge, positive streamers originating from the upper poles of the drops may travel upwards to neutralize negatively charged hydrometeors and so lower some of the main negative charge. Meanwhile, negative streamers from the lower poles of the drops may initiate the downward moving stepped-leader stroke of the lightning discharge that, once started, may be propagated and sustained by a growing wave of potential gradient travelling through a medium composed of ions and charged droplets produced by the filamentary drop discharges. However, the details of the process have still to be worked out.

In summary

On the basis of the observations, experiments and calculations described in this chapter, it is concluded that the primary mechanism of charge generation and separation in the thunderstorm essentially depends on the formation, growth and electrification of pellets of soft hail. These become polarized in the positive fine-weather electric field and then cloud droplets and ice crystals rebounding from their undersurfaces carry away some of the induced positive charge and leave the hail pellets negatively charged. Gravitational separation of the pellets from the much smaller positively charged particles intensifies the electric field, and the whole process continues to accelerate until the electrical forces on the charged hydrometeors oppose further vertical separation. The large-scale field then reaches a limiting maximum value of about 4000 V/cm in a time determined largely by the precipitation intensity. It appears that rather stronger fields than this are required to cause electrical breakdown of the air and initiate a lightning discharge, even in the presence of large raindrops, but such fields may occur locally where two oppositely charged regions are contiguous as in the base of most thunderstorms.

Laboratory experiments suggest that charges may also be separated during the melting of ice, the freezing of supercooled droplets, and the disintegration of raindrops, but these mechanisms appear to be of only secondary importance in thunderclouds. Electrification is produced by the bursting of air bubbles in melting ice, by the breakup of large raindrops, and their splashing on the ground, because strong shearing stresses are set up at the water–air interface and cause rupture of the electrical double layer, but this basic mechanism is overwhelmed by the much larger charges induced on the hydrometeors by the primary thunderstorm field. Of course both the electrical double layer and the induced charges have their origins in the dipolar properties of the water molecule.

Accordingly, it appears that the growth of the primary field, the formation of the lower positive charge, and the initiation of lightning discharge are all largely due to the inductive charging of hydrometeors, in the first place by the fine-weather field. Consequently, if the direction of the latter were reversed, the signs of the inductive charging processes and hence the polarity of the thundercloud would be reversed, but the underlying physical mechanisms would not be affected.

Although our conclusions appear to be entirely consistent with the observed structure and behaviour of thunderstorms, they rest very heavily on the results of laboratory experiments and theoretical calculations which can never completely simulate developments in a natural cloud. This sets a premium on more reliable and representative measurements of electric fields, currents, and charge distributions in thunderstorms, and of the charges carried by hydrometeors, all in relation to the type and intensity and of the precipitation. Such information will not be easily acquired, not only because of the difficulty and danger of flying aircraft into the actual centres of thunderstorms, but because the aircraft itself is likely to distort the electric field, produce corona discharges, and so make the measurements uncertain and difficult to interpret. However, a useful start could be made by studying the growth of both the electric field and the precipitation in the early stages of an incipient storm. Thus a demonstration that the growth of the electric field in a shower-cloud is closely correlated with the formation and growth of soft hail, that the hail pellets are negatively charged, and that only much weaker fields occur in *rain*-showers of much the same intensity, would constitute strong evidence for our favoured mechanism, even though it might not be possible to follow developments right through to the lightning stage.

APPENDIX

Collision efficiencies for drops of radius R colliding with droplets of radius r at 0 °C and 900 mb

$R(\mu)$	2	3	4	6	8	10	15	20
10	0·013	0·020	0·028	0·036	0·037	0·027	—	—
20	—	—	0·015	0·030	0·052	0·072	0·069	0·027
30	—	—	—	0·040	0·11	0·23	0·54	0·56
40	—	—	—	0·19	0·35	0·45	0·60	0·65
60	—	—	0·05	0·22	0·42	0·56	0·73	0·80
80	—	—	0·18	0·35	0·50	0·62	0·78	0·85
100	0·03	0·07	0·17	0·41	0·58	0·69	0·82	0·88
150	0·07	0·13	0·27	0·48	0·65	0·73	0·84	0·91
200	0·10	0·20	0·34	0·58	0·70	0·78	0·88	0·92
300	0·15	0·31	0·44	0·65	0·75	0·83	0·96	0·91
400	0·17	0·37	0·50	0·70	0·81	0·87	0·93	0·96
600	0·17	0·40	0·54	0·72	0·83	0·88	0·94	0·98
1000	0·15	0·37	0·52	0·74	0·82	0·88	0·94	0·98
1400	0·11	0·34	0·49	0·71	0·83	0·88	0·94	0·95
1800	0·08	0·29	0·45	0·68	0·80	0·86	0·96	0·94
2400	0·04	0·22	0·39	0·62	0·75	0·83	0·92	0·96
3000	0·02	0·16	0·33	0·55	0·71	0·81	0·90	0·94

The column headings are grouped under $r(\mu)$.

INDEX